Academic Writing for Publications

Zhongchao Tan

Academic Writing for Engineering Publications

A Guide for Non-native English Speakers

 Springer

Zhongchao Tan
Department of Mechanical
and Mechatronics Engineering
University of Waterloo
Waterloo, ON, Canada

ISBN 978-3-030-99366-5   ISBN 978-3-030-99364-1   (eBook)
https://doi.org/10.1007/978-3-030-99364-1

This Springer imprint is published by the registered company Springer Nature Switzerland AG
The registered company address is: Gewerbestrasse 11, 6330 Cham, Switzerland

To my Mother

Acknowledgments

I would like to thank the publishers for their permission to use the materials cited in this book, as well as those giving me permission to reprint materials.

Contents

Part I

Introduction

Introduction

1.1 About This Book

Writing skills are important to effective communication. Written communication is more than fluent speaking or a good command of grammar, spelling, and punctuation. Writing is a complex task that requires systematic training and continual practice of many techniques, such as organizing ideas logically, constructing sentences and paragraphs coherently, presenting with appropriate tones, formatting in a stylish manner, and executing in an ethical way.

Every engineer is a writer. Written communication reveals the author's intelligence of thinking, ability of using words, level of education, and so forth. Good writers are usually creative people with brilliant ideas, which can help capable writers excel in their career development. Regardless of your job, you might need to write every day. Writing is a great way to extend your voice and convey your thoughts and ideas to many people in the world.

Academic writing is also referred to as scholarly writing. It is the writing produced as part of academic works, such as an article, a book, a report, a thesis, and the like. Academic works, which are primarily produced by researchers and graduate students, are shared with other professionals in close areas of research. Engineering academic writing is one type of technical writing produced by authors in engineering.

This book is aimed at non-native English writers such as international students, as well as researchers who are studying and working in English-speaking countries. This book uses hundreds of examples for contrast and comparison, many pairs of confusing words, and an emphasis on some essential cultural difference for non-native English writers. Furthermore, an introduction of copyright and plagiarism at the beginning can help writers avoid unnecessary challenges from their readers or the copyright holders.

1.2 Scope and Readers

Writing in English can be formal or informal. Formal writing in English is used by students in academics, professors, professionals in engineering industries, etc. It is expected to be professional and should follow the rules of grammar, spelling, logic, *etc*. However, informal writing does not have to rigorously follow grammatical criteria, and it is often for local use by a particular ethnic or social

© The Author(s), under exclusive license to Springer Nature Switzerland AG 2022
Z. Tan, *Academic Writing for Engineering Publications*,
https://doi.org/10.1007/978-3-030-99364-1_1

group in a casual setting. Most informal writing tends to be vivid, colorful, and impressive, but its usefulness is limited to certain contexts because of its use of dialects, idioms, jargons, or slangs.

This book focuses on formal academic writing in a professional language and frame. It is written in standard English and provides useful guidelines on development of thoughts, organization of ideas, construction of paragraphs and sentences, and choices of precise words. It also pays attention to details such as visuals, punctuation, and format. Informal writing is excluded from the scope of this practical guideline.

This book focuses on the following two parts that are challenging to many engineering writers:

- **Organizing ideas:** thinking like an engineer, which covers organization of ideas in engineering writing for scholarly publications
- **Language skills**: using engineering languages to write with clarity and conciseness, which covers the basics of engineering writing for scholarly publications

This book is designed for non-native English speakers who need to write scientific research articles, technical reports, proposals, engineering thesis, academic books, and other technical documents in English. Admittedly, many books are available for the training of technical writing; some are free online, and others are priced for sale. However, few of them are dedicated to non-native speakers of English for academic writing. Most international students in English-speaking countries are capable of reading and understanding published works. As an engineering professor, however, I have seen and helped many students who need systematic training in writing for their course projects, conference papers, journal articles, theses, and the like. The guidelines in this book have been proven useful to advancing their formal writing skills.

Readers of this book are not limited to non-native English speakers. Students can use the book as self-study materials to improve their communication skills, and engineering instructors can use the book for writing skills development. Engineering students may find this book well balanced between complicated theory and simplicity. Researchers and professionals in engineering industries may find valuable writing techniques in the book too. It is useful to new or experienced writers. With the guidelines in this book, they both can further convey complex technical information with clarity and conciseness.

Nonetheless, this book is not focused on general English writing, business writing, or fiction. Someone who dreams of becoming a novelist certainly should not use this book either.

1.3   Organization of This Book

In this book, the approach to writing is presented in the order of time following a typical writing sequence. A normal sequence in academic writing for engineering publication is preparation, organization, and language refinement:

1. Preparation for purpose, scope, readers, etc.
2. Outline for structure and logics
3. First rough draft for ideals organization
4. Multiple revisions for clarity, coherence, and conciseness
5. Proofreading for language refinement
6. Preparation for publication

Fig. 1.1 General procedure of writing

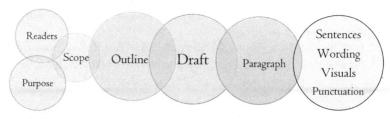

1. Preparation   2. Organization of ideas   3. Language

Organization of ideas precedes language refinement. Successful writing starts with preparation and outline, continues with draft and revision, followed by language refinement, and ends with formatting and proofreading. These steps often overlap without a clear boundary. One step upstream may affect subsequent steps. For example, you may expand the scope of the work when you are working on *Results and Discussion*. It is difficult to quantify the time required for each step, but in general more time is needed for a writing task with a greater complexity. Dividing the writing process into different steps may also be important to collaborative writing for multiple authors. However, collaborative writing is out of the scope of this book.

Figure 1.1 illustrates the general procedure of writing. Accordingly, this book is organized into three parts, which may overlap in practice. For example, outline helps with preparation and paragraphing contributes to organization of ideas.

Part I. Introduction
It includes Chap. 1 (this chapter) and Chap. 2, which covers academic integrity and basic rules in engineering writing.

Part II. Organization of Ideas
This part includes Chaps. 3, 4, 5, 6, 7, 8, 9, and 10: Preparation, Outline, Introduction, Methodology, Results and Discussion, Conclusion, and Other Elements in a typical engineering article or thesis.

Part III. Language Skills
This part is focused on intermediate and advanced language skills that help you improve the clarity and conciseness of your writing. They are presented from Chaps. 11, 12, 13, 14, 15, 16, and 17: Clarity and Conciseness, Paragraphs, Sentences, Phrases and Words, Visuals, Punctuations, and Finalization. The rest of the book geos to front matter and back matters such as references, appendices, and index.

Part IV. Correspondence with Editors
This is a short section with only one chapter. It emphasizes the importance of scope and readers introduced in Preparation stage of writing and the essence of clarity and conciseness in response to reviewers' comments.

Part I prepares you for writing. It introduces academic integrity and professionalism, which are critical to publications that may affect countless readers in the world.

Part II focuses on ideas organization. *Preparation* and *Outline* are the first steps in writing. You need to keep your scope, purpose, and readers in mind throughout your writing. A well-developed outline sets up the main stage for writing. Drafting follows the outline with organized paragraphs. The first draft of a typical engineering publication (e.g., a journal article or a degree thesis) is composed of multiple sections, including *Introduction*, *Methodology*, *Results and Discussion*, and *Conclusions*.

Each of them follows an engineering-specific structure, logics, and rules of presentation. They are introduced in this part.

Part III is for the revision of draft produced in Part II. Revisions follow the rough first draft; when you revise the draft manuscript, pay attention to paragraph structures, sentence structures, tenses, tones, voices, phrases, words, etc. This part also includes the rules of punctuation and grammar. Revisions are time-consuming, but you must revise your document again and again until you are satisfied before you share your work with other professionals.

The last steps finalize your manuscript. They include proofreading and communication. This section also covers the best practices that help you avoid challenges in plagiarism and copyright.

Part IV, *Correspondence with the Editor*, is rarely available in similar books, but it is a critical part of academic writing for engineering publications. After all, it is the editor, with the input from peer reviewers, who decides the acceptance of your manuscript. This part guides you in professional, clear, and concise correspondence with the editor who handles your submission.

To be practical and useful to the intended readers, theoretical writing skills are presented for the sake of simplicity with examples. Many examples compare correct writing with incorrect ones from different sources. The examples, which are blocked and numbered in sequence, stand out from the text. This layout allows intermediate writers to quickly improve writing skills by focusing on the examples.

1.4   American and British Spellings

American English is used as much as possible in this book, but British English may be used unintentionally. American and British English spellings are different; either one of them should be consistently used in the same document. Admittedly, it is a challenge to non-native English speakers who speak neither of them to tell the difference, but it is a mistake to mix them in one single document. Appendix A2 explains some common differences between American English and British English (Williams 2018; Oxford 2021; Tysto 2021).

Canadian English is a mix of American and British spellings. Although most of them are British-like, some of Canadian spellings are American-like. For example, *tire* instead of *tyre* is used in Canadian English. For some words, Canadians accept both British and American spellings. Nonetheless, Canadian English is not used in academic writing, especially peer-reviewed journal articles. Therefore, Canadian students are encouraged to learn American or British English conventions (Curtsey JP Hickey).

Academic Integrity

2

2.1 Plagiarism

Plagiarism is prohibited in many sectors, including higher education, engineering research, and professional publication. Plagiarism may lead to serious professional and legal consequences for the individuals who plagiarize. Their employers are often impacted too.

In technical writing, you must give credit to the original creators of the unique ideas. You can do so by quotation and paraphrasing with appropriate in-text citations and references. Paraphrased materials should also be credited if the unique ideas belong to others, even when it is only one line. You also need written permission, with or without payment, from the copyright holders (*see* Copyright).

Software is available for plagiarism checking. For example, the University of Waterloo in Canada recommends Turnitin and iThenticate for plagiarism checking of student work and academic publications.

2.2 Quotations and Paraphrasing

You can avoid plagiarism by direct quotation, indirect quotation, and paraphrasing. For any of these, you also need to use references and in-text citations. (*See also* In-Text Citation and References.) They are introduced as follows.

You can occasionally use a direct or indirect quotation to stimulate interest in your subject. You may use quotations for drafting but use as few quotations as possible in the final version of your works. It is better to rephrase the direct quotations in your own language and give credit to the original creator.

Indirect quotation is usually introduced by the word *that*. However, indirect quotation may lead to wordiness because of the names preceding the word *that*. In addition, technical writing emphasizes the ideas rather than the creators of the ideas. Indirect quotations appear to be your own writing, but not. You must provide in-text citations and references to credit the creators of the original ideas. For all these reasons, paragraphing is a better option than quotations, direct or indirect.

Paraphrasing is representing the essential ideas that belong to others using your own language; it is also called *rewriting*. Paraphrasing is a key skill in academic writing because it saves your reader's time by synthesizing the knowledge in the literature. Example 2.1 compares direction quotation, indirect quotation, and paraphrasing.

© The Author(s), under exclusive license to Springer Nature Switzerland AG 2022
Z. Tan, *Academic Writing for Engineering Publications*,
https://doi.org/10.1007/978-3-030-99364-1_2

Example 2.1. Quotations and Paraphrasing

Direct quotation: Stephen Hawking once said in [*year*], "Quiet people have the loudest minds."

Indirect quotation: Stephen Hawking (*year*) said that quiet people have the loudest minds.

Paraphrasing: Quiet people normally have the loudest minds (*Hawking, year*).

Example 2.2. Paraphrasing Earlier Work with Citation

- **Original writing:**
  "Aerosol particles are important components of the atmosphere and strongly influence the Earth's climate by directly scattering or absorbing sunlight and indirectly by aerosol–cloud interactions. Aerosol particles are also involved in negative health effects caused by air pollution and are linked to increases in respiratory and cardiovascular diseases. All these particle effects are influenced by the chemical composition of the aerosol particles...." (Fuller et al. 2012)
- **Paraphrasing:**
  Atmospheric aerosol particles strongly affect the climate on the planet earth because they can directly scatter or absorb sunlight and indirectly interact with the cloud. They also negatively impact human health by air pollution, causing respiratory and cardiovascular diseases. The effects of aerosol particles are attributed to their chemical composition (Fuller et al. 2012) ....

Paraphrasing common knowledge does not need a reference or citation. Common knowledge is widely known to the public or understood by your target readers. For example, Newton's second law is common knowledge and is found in all creditable engineering dynamics textbooks. If you are not sure, however, citations and references protect you.

Again, you must provide in-text citation and reference for the text paraphrased from any earlier publications. Example 2.2 shows how to paraphrase the ideas in an earlier publication. Both paragraphs have similar ideas, but they are written with different languages.

Note

Copy and paste the contents from your own earlier publications may also result in plagiarism because you may have transferred the copyright to the publishers.

2.3   Copyright

Materials, especially visuals, under the protection of copyrights cannot be published without written permission from the copyright holders. You need to contact the copyright owners for permission and keep the written permission for future record.

Copyright holders are not necessarily the creators. For example, most authors sign legal agreements to transfer their rights to the publishers of their journal articles. In this case, the publishers own the copyrights. You may request and receive permissions to use the materials from the copyright holders, but you still need to give credit to the original creators of the ideas. *See* Plagiarism.

There are multiple ways to request copyright permissions. You can contact the author(s) directly or request through Copyright Clearance Center (copyright.com). For example, I requested through the Copyright Clearance Center and received the permission from the American Chemical Society (ACS) to use a large body of the sample article by *Fuller* et al. (2012) for examples in this book.

Nowadays, many online materials are available to support open access. In this case, you can take a snapshot as a proof. Otherwise, request for permission following the instruction on the website.

2.4   Privacy and Confidentiality

Private information must be protected according to privacy standards in some countries. Ensure that your writing for publication follows both local and international regulations. Consult with your superior if you are in doubt.

Photographs are often used to show the appearance and the size of an object. Without permission from the copyright holders, however, they cannot disclose the confidential working mechanism. Such details are normally represented in a schematic or drawing. A permission to use also helps avoid plagiarism by giving credit to the photographers.

2.5   Authorship

Authorship has become unnecessarily complicated with the growing demand for multiple disciplinary research collaboration. In some countries, only the first author or the contact author(s) merit career advancement, while in others, all co-authors are considered equally important to the work. The number of publications may impact their career advancement (such as tenure and promotion) or personal incomes (e.g., grants and bonuses).

Example 2.3. Authorship

J Cao, S C Lee, J C Chow, J G Watson, K F Ho, R J Zhang, Z D Jin, Z X Shen, G C Chen, Y M Kang, S C Zou, L Z Zhang, S H Qi, M H Dai, Y Cheng, K Hu, 2017. Spatial and seasonal distributions of carbonaceous aerosols over China, *Journal of Geophysical Research* 112: D22S11; doi:10.1029/2006JD008205.

No single definition of *authorship* applies to the world. It is up to the co-authors to withhold their academic integrity and responsibility of the published works. It is important to understand the difference between authors and contributors. Incorrect authorship often treats contributors as co-authors.

The contributors may have enabled the research by providing valuable resources or administrative support, but they do not participate in the research work or writing.

On the other hand, it is ethical to add co-authors who contributed to the merit of the work. Most journal articles, technical reports, and similar works are written by teams with complementary expertise, whereas a degree thesis is likely by an individual who receives the degree. Regardless of the publication, the co-authors should be the writers, investigators, or team members who contribute to the knowledge production. The contributions can be active involvement or supervision of one or more of the following activities:

- Concept design
- Methodology development
- Data collection
- Data analysis
- Manuscript drafting
- Manuscript revision
- Approval of final version for publication
- Ensuring accuracy and integrity of the work

Take authorship seriously and avoid free riders in authorship. Average authorship has grown from three or four co-authors to six or more over the last 20 years. It has caused confusion over accountability and entitlement among academics and professionals. Regardless of the motivations, an unnecessarily long list of co-authors calls for negative perception. Listing too many co-authors dilutes the contribution of each and appears unreasonable, especially to the fellow professionals who understand the efforts it takes to produce certain type of work.

2.6   Logical Writing

The goal of academic publication is sharing knowledge for the greater good of humankinds. The ideas shall be clear to readers and easy to understand. Thus, expository writing is required for engineering publication. It enhances accuracy, clarity, and conciseness of writing. You must have a thorough understanding of the subject to effectively use expository writing. Do not leave it to your readers to analyze and draw conclusions based on limited information.

Logic is essential to convincing arguments and valid conclusions. Drawing an illogical conclusion is considered unethical because it misleads the readers. Illogical writing undermines the writers' credibility and may have a negative impact on their professional career development. Examples of illogical writing are as follows:

- A statement contrary to common sense. It calls for reasoning. (If you write down "a helium balloon falls down," then this sentence should be followed by a good explanation.)
- An overinclusive statement. It generalizes an observation which is applicable to only a small group. (Some students tend to claim that their computational models are the *best*.)
- A mis-linked statement. It is connected to a previous statement with a logic gap.
- A questionable cause. A logical error results from hasty conclusion without examining the cause. (For example, I felt stomachache after eating at a restaurant, so the food in that restaurant must have made me sick.)

2.7 Unbiased Writing

Academic writing for engineering publication has its own style beyond an individual's personal and distinct stylistic traits. This applies to engineering and scientific articles, technical books, technical encyclopedia, master's theses, and doctoral dissertations. Therefore, the writer's tone should be impersonal and objective, rather than emotional and subjective, for the greater benefit of society instead of the writer's personal interest.

Thus, you must avoid deceptive or confusing language in your writing. It is unethical to deceive or mislead your readers with ambiguous words with multiple meanings or misleading information.

Never make false claims. To draw a conclusion from omitted or incomplete data, questionable sources, or selected evidence is not only illogical but also unethical. It may be even illegal to report false, fabricated, or plagiarized results in engineering publications.

Avoid drawing a conclusion from partial information if you know there is more. For example, drawing a conclusion with selected experimental results is misleading because that is valid only under a narrow range of conditions. It is your obligation to correct any misrepresentations of fact before, and after, publication. Make sure you can support your argument with reasoning and evidence. The following techniques may help you avoid biased writing:

- Present evidence, including verifiable data or statements from creditable sources, instead of subjective opinions. (*See also* 5.2.) Make sure that the data support your statement. Allow your readers to validate your conclusions from the facts, statistical analyses, and examples that you present.
- Present ideas with controlled pace to your readers. Assume your readers are new to the subject, although you might be the expert.
- Present negative facts or conclusions as they are. Understanding the limitations of your work indicates your qualification and expertise in the field of research. If some data collected appears to be against your statement, you still must present them and explain why they are excluded from data analysis.

Visuals like graphs, images, and even color require careful attention. Make sure that visuals are presented with professionalism. Always create clear simple visuals with consistent labels in the same writing. Avoid unnecessary complication. Visuals can also be misleading when information is selectively omitted or distorted. Distortion can be created by incorrect scales or selected data. *See* Chap. 14. Visuals for more information.

2.8 Neutral Language

We live in a modern world that respects different races, genders, religions, and so on. In addition to cultural diversity, modern society calls for inclusivity. Words matter to ethnic groups, nationalities, religions, and other identifying groups. You should use politically neutral words and tones to avoid misunderstanding.

Acknowledgment of diversity and recognition of cultural difference are key to effective communication. It is essential for a non-native speaker of English to recognize that the readers may come from a variety of cultural backgrounds anywhere in the world. What is considered acceptable or efficient in one language might be vague or blunt in another and vice versa.

Thus, it is important to avoid offensive language and choose common words that are acceptable to as many people as possible. Indeed, writing for international readers posts a challenge to you, but it offers the opportunities for you to reach more readers and to create a strong impact on society.

Gender inclusivity also affects grammatical agreement. Personal pronouns must agree in gender with their antecedents, which are the nouns they refer to. Agreement in gender for plural pronouns is not an issue because they automatically agree with antecedents. For singular pronouns (*he/him/his, she/her/hers*, and *it/its*), however, gender agreement may be challenging to non-native English writers because singular pronouns are not gender specific in some languages (e.g., Chinese). In English, the singular pronouns are gender specific, and you need to match them with the genders of their antecedent.

Genders other than male and female have been recognized for centuries in many nations and religions, even in traditional English-speaking countries. With growing awareness of gender inclusivity, several countries have passed laws recognizing sex identities other than the male or female bodies. You can consider the following approaches to gender inclusivity in your writing:

1. Without sacrificing clarity, use indefinite pronouns (*everyone, anyone, person, individual, each*, etc.) and nouns (*member, student, performer, child, person*, etc.). They apply to both males and females.
2. When you must, use *he or she, him or her*, and *his or her*, especially when the gender is unidentified.

The combinations of <u>*he or she*</u>, <u>*his or her*</u>, and so on impact the smooth flow of reading if repeated in one sentence. Alternatively, you can take any of the following two options:

• Use *they/their/theirs* with a plural antecedent.
• Rewrite the sentence to remove the pronoun.

Example 2.4. Gender Inclusivity in Writing
• He or she introduced the speaker to the audience.
• Every author has provided his or her email address.
• All authors have provided their email addresses.
• The moderator introduced the speaker to the audience. *[This is rewording.]*

Note

Avoid gender-related bias. For example, matching *the nurse* or *the engineer* with *she* or *he, respectively,* without confirmation is considered a bias in English writing. Both male and female can be nurses or engineers in many countries.

There are many other writing techniques that have cultural implication, for example, active voice, positive tone, direct statement, and neutral color. They are emphasized as they appear in the text throughout the book. In short, keep your international readers in mind and write with cultural inclusivity.

2.9   Clear and Concise Writing

Engineering publications emphasize *clarity* and *conciseness*. A good understanding of grammar and special terminology enables you to communicate clearly and precisely. However, academic writing for engineering publications takes much more than grammar and spelling. Engineering writing is characterized with clarity and conciseness from the beginning to the end of the entire document.

Your vocabulary ought to be specialized and precise instead of fancy, elegant, or dull. Therefore, contractions, slangs, or dialects are *not* used in the academic writing. The key is to keep your readers in mind; they are from diverse cultural backgrounds with different levels of education and experience. Ask yourself this question while you are writing: what is important to the readers? Furthermore, intensifiers (*most*, *much*, *very*) should be used for emphasis with care – often they are unnecessary in engineering academic writing, which values objective data and logical argument.

Meanwhile, professional writing can be interesting and lively too. Using active voice, positive tone, various structures, balanced emphasis, and subordination reduces monotonous writing style (*see* Example 10.9).

In addition, communicating with goodwill and modesty is preferred in academic publications (*see* Example 2.5). For a publication aimed at international readers, who may be foreign to the subject, you need to elaborate with sufficient details. In addition, modesty is more acceptable than arrogance in any formal writing.

There are numerous skills to improve clarity and conciseness in your writing. You can find related techniques and examples throughout this book. Further, you need to keep your readers in mind: ensure that they can understand you solely by reading your works.

Example 2.5. Writing with Goodwill and Modesty
Arrogant: Our device can capture particles at a <u>100% of efficiency</u>.
Modest: <u>Based on the results</u> herein, we can say that our device can capture particles at an efficiency of <u>almost 100%</u>.

The rest of this book introduces the techniques for clear and concise writing. Clarity requires both logical organization of ideas and skillful use of language. They are at the core of this book: Part II focuses on organization of ideas, and Part III, on language skills. Keep clarity and conciseness in mind while reading this book and writing in the future. Nonetheless, you need to practice writing as much as possible in a technical context. It takes time and continual effort to become an effective writer.

2.10   Practice Problems

Question 1 (plagiarism check) login to iThenticate and practice plagiarism checking using one of your earlier publications (https://www.ithenticate.com/).

Question 2 (copyright) login to Copyright Clearance Center and request a permission to use one of your earlier publications. It is normally free of charge (https://www.copyright.com/).

Question 3 (political correctness) Revise these sentences to improve gender inclusivity.

1. Tan (2004) developed an analytical model for uniflow cyclone, and he validated his model using his own experimental data.
2. The data reported by Chang (2021) are inaccurate; she did not consider the surface effect in her model.

Question 4 (paraphrasing) Paraphrase the following text (source: Fuller et al. 2012).

1. The composition also depends on the size fraction of the aerosol.
2. The analysis of the organic fraction of atmospheric aerosol particles on a molecular level is often challenging.
3. The analysis of the organic fraction of atmospheric aerosol particles on a molecular level is often challenging because their composition can be highly variable in time and space and usually only small sample amounts are available for analysis because atmospheric concentrations are typically a few micrograms per cubic meter. The composition also depends on the size fraction of the aerosol; for example, resuspended, wind-blown particles are mostly larger than 1 μm and have often a very distinct chemical composition compared to particles smaller than 1 μm, which are mainly emitted by combustion sources or formed by chemical reactions in the atmosphere.
4. We have developed a new technique to analyze the organic composition of atmospheric aerosol on a molecular level with a high time resolution by combining a new extraction and ionization technique, LESA, with RDI samples and ultrahigh- resolution mass spectrometry. LESA facilitates analysis without time-consuming sample preparation or extraction and also allows direct extraction from a sample surface with high spatial resolution, which when combined with RDI samples results in highly time-resolved information of ambient aerosol organic content.

Part II

Organization of Ideas

3.1 Purpose of Writing

Any academic writing for publication has a purpose. Academic publications include articles, books, reports, and theses. A typical research article is thousands of words long. Peer-reviewed research articles are published in academic journals, which are often referred to as *periodicals* in a library. A book comprises multiple chapters with a coherent focus. Conference papers are short articles written for the conference, where you can present your results to the research community. A conference paper is normally shorter than a research article. A thesis is a degree-requirement work written by a research-based graduate student. It takes years for a doctoral student to complete a dissertation. Although they have different sections, most of the sections have common organization and structures. Conference papers and peer-reviewed journal articles may be based on the contents of dissertation.

Although the guidance in this book is aimed at academic writing, the basic writing principles often apply to similar documents such as progress reports, lab reports, investigative reports, test reports, white papers, and course project reports.

The two most important academic writing applications in engineering are journal articles and theses. Journal articles published in engineering periodicals aim to further the knowledge and to report the advances in a specialized field. The readers are normally professionals, such as academics, educators, engineers, graduate students, and scientists, who also regularly contribute articles to those journals. Research articles published in journals, especially those prominent international journals, may raise the profile of your employer and improve your opportunity for professional advancement.

There are several golden standards for publications in prominent international journals. First, your work (knowledge or approaches) must be original. Second, the significance of your contributions justifies the time and effort to write an article as well as the time of others for them to review and publish your knowledge. Other factors such as the prominence of the journal, review time, and frequency of citation are also important before you begin writing.

3.2 Identifying Readers

It is important to identify your readers before writing. Ask yourself the following questions before writing down anything. Precise answers to these questions will also determine your scope of writing and levels of clarity and conciseness in your writing (Alred et al. 2018):

© The Author(s), under exclusive license to Springer Nature Switzerland AG 2022   17
Z. Tan, *Academic Writing for Engineering Publications*,
https://doi.org/10.1007/978-3-030-99364-1_3

- How much background information do I need to put down in writing?
- Are my readers qualified professionals in the scope of the work?
- What do my readers already know about the subject?
- Should I define basic terminology?
- Am I communicating with international readers?

Keeping your readers in mind is crucial to effective communication with them. Consider your readers' levels of knowledge. Many publications are for readers with diverse backgrounds. You need to accommodate their needs as much as you can. Most engineering publications are aimed at international readers worldwide. Admittedly, it is challenging to satisfy all the readers with diverse needs. Then, you should determine your primary readers, such as the students in your professional community, your peer researchers in your research areas, or those who may make decisions based on your writing.

In addition, you ought to understand their cultural values that underlie the language used for international readers. In a globalized world like today, you are urged to write with clear and complete sentences. You should avoid the following writing mistakes, which may confuse international readers:

- Abbreviation, acronym, and terminology without definition first
- Overly simplified style
- Unusual word order or rambling sentences
- Informal words for humor, irony, and sarcasm
- Localism, jargon, slangs, and idioms

Regardless of your readers, any clear writing in English is expected to be concise and coherent. Typical academic writing is for educated readers, including university students, researchers, and professionals in specialized industries. These highly educated readers determine how you write, and you need to make the following decisions accordingly:

- Provide headings and subheadings using words in the field of primary readers.
- Provide enough background information with details.
- Use neutral and technical vocabulary in body text.
- Write a procedure with details so that readers can follow easily.
- Use clear visuals besides written text for clarity.

3.3   Scope of Writing

The scope should be determined before drafting the document, although you may refine it at the later stages such as drafting and revisions. A well-defined scope saves you time in writing. The work you are writing should match the scope of the journal to be submitted for consideration of publication. (*See* Cover letter.)

Effective writing has a clear focus. Avoid overly broad objectives. Simply ask yourself what your contributions, and only yours, are; then you can establish your primary scope of writing. Be precise when you state the objectives, which are normally justified by state-of-the-art background review. *See also* Drafting Objectives.

3.4   Structure of Writing

The structure of engineering academic writing depends on the length of the document. The two most important types of academic writing are research articles (short) and theses or books (long). Although they follow similar principle and writing styles, their structures are different. Even the structure of journal article varies with journal. Their common features are introduced as follows.

Table 3.1 shows typical divisions that technical scholarly writing may have in the order they typically appear. Book and thesis structures are normally more flexible than those of conference papers and journal articles. In addition, most periodicals and conference organizers use their own prescribed structures for the major sections defined by level-one headings. Almost all journals furnish guidelines for authors and make them accessible on the journal websites.

The common level-one headings in most international journals include *Title*, *Authorship*, *Abstract*, *Introduction*, *Methodology*, *Results and Discussion*, *Conclusions*, and *References*. Optional but valuable sections include *Recommendations*, *Limitations*, and *Acknowledgments*. Some journals may include *Supplemental Materials*, which are unavailable in printed hardcopy but are accessible online.

However, it does not mean they all will appear in one single document. Long documents such as books and theses normally have more sections than short documents do. You need to check with the publisher for the prescribed sections and ensure that you have the freedom to create your own sections and subsections.

Despite the variation of sections in academic writing, they share the same writing principles. They are introduced as follows.

Table 3.1 Typical sections in engineering publications

Group	Section (Heading 1)	Long document	Short document
Front matter	Title	Yes	Yes
	Authorship/Affiliation	Yes	Yes
	Abstract	No	Yes
	Executive Summary	Yes	No
	Table of Contents	Yes	No
	List of Figures	Yes	No
	List of Tables	Yes	No
	Foreword	Optional	No
	Preface	Optional	No
	List of Abbreviations, Acronyms, Symbols	Yes	Optional
Body	Introduction	Yes	Yes
	Chapters	Yes	No
	Methodology	Yes	Yes
	Results	Yes	Yes
	Discussion	Yes	Yes
	Conclusions	Yes	Yes
	Recommendations	Yes	Optional
	Explanatory Notes	Optional	No
	Acknowledgments	As needed	As needed
	References	Yes	Yes
Back matter	Appendices	Yes	As needed
	Bibliography	As needed	No
	Glossary	As needed	As needed
	Index	Optional	No
	Supplemental Materials	No	As needed

3.4.1   Front Matter

Front matter describes the general ideas of the work and summarizes the sections of the document. Although not every academic writing includes all the elements of front matter listed in Table 3.1, *Title*, *Authorship*, and *Affiliation* are essential to all documents. The rest are optional. For example, *Abstract* is usually mandatory for journal articles and dissertations, and *Foreword* or *Preface* is essential to books.

3.4.2   Body

A document body contains the background information (*Introduction*) for the work, current state of the research, procedures for conducting the research, results, discussion, and conclusions drawn from the results. Recommendations and limitations of the work further enhance the value of writing, if appropriate. Furthermore, all this information can be enhanced by visuals, including figures and tables.

Introduction catches the interest of your readers. It explains the current state of the work, the knowledge gaps to be addressed, the questions to be answered, and the problems to be solved through *your* research. The *Introduction* also answers two essential questions: why your contributions are important and new. It typically ends with well-defined objectives.

Methodology usually follows the statement of objectives. This section describes the procedures of obtaining your own data, if any, in addition to data in literature. *Methodology* should be written with sufficient details (as if it were a manual), allowing readers to replicate the data. At the minimum, readers have the right to know the experimental procedure, raw data collection method, and expected results to support the conclusions.

Results and Discussion are presented after *Methodology*, and *Conclusion*s summarize the key contributions of your work. Avoid biased opinions toward your own conclusions. In attempt to establish a logical argument, you also need to draw information and data from external sources. They must be credited by references and in-text citations to avoid plagiarism. The references are listed after the body text. Visuals reinforce your argument, and they should be presented with appropriate styles. (*See* 14. Visuals.)

3.4.3   Back Matter

Back Matter contains additional and supplementary materials, which are useful to the readers for various reasons. *Bibliography*, which is different from *References*, provides the readers with additional information that concerns most readers. *Bibliographic* entries may not be cited in the text; they are simply listed for additional reading. *Appendices* (for long documents) or *Supplemental Materials* (for some journal articles) expand on the subjects that might distract the primary readers, if they were included in the main body, but they are useful to others who are interested in a better understanding of the subjects. *Glossary* and *indices* are for multi-chapter and multi-volume documents such as books, dissertations, and encyclopedia. They are useful tools for the readers to locate relevant information in the main body. (*See also* Chap. 9 for more information.)

3.4.4   Headers and Footers

Headers appear at the top of designated pages; footers appear at the bottom of the pages. The text in headers and footers varies with publication. The headers may include the document title and the section title, and the footers contain other information to help readers track their reading progress. Either the headers or the footers should include the page numbers, more often as footers than headers. You certainly can be creative in your use of headers and footers, but you should avoid crowding them. Finally, information in headers and footers should be concise, and it is not meant to distract readers from the main text.

Outlining

4

4.1 Outlining Steps

You normally have few opportunities to observe how outlines are created and revised by reading the final version of the documents. Many engineering programs do not offer academic writing courses to graduate students; most of them are trained by their academic supervisors. As a result, academic writing skills become personal and are often omitted from engineering education.

The Outline is meant to make your writing easy, but it should not dictate your flow of ideas or choices of words. An outline is only a nonrestrictive guideline that guides your drafting. Like a road map for driving, the outline helps you focus on the purpose of writing so that you can logically draw your conclusion. Nonetheless, you can always change when you need to.

There are three nondistinct steps for the creation of outlines:

1. Division with headings
2. Short topic outline with keywords
3. Lengthy outline with sentences

The first step divides your document with key headings that reflect the sequence of presentation. The second step details the flow of ideas with short phrases and keywords. The last step extends short phrases and keywords into complete sentences; each sentence in the outline states the controlling idea of the corresponding paragraph in the draft.

Step 1: List Major Headings
Typical academic writing for engineering publications begins with introduction, followed by methodology, continues with results, and ends with convincing conclusions. Accordingly, the first step of the outline begins with headings, lists, and other special design features to frame for your writing. At this point, you should also consider the title that defines the scope of your writing. There are normally more than four Level-1 headings. For example, the major headings of a typical international journal article comprise of at least five levels as follows.

© The Author(s), under exclusive license to Springer Nature Switzerland AG 2022
Z. Tan, *Academic Writing for Engineering Publications*,
https://doi.org/10.1007/978-3-030-99364-1_4

Example 4.1. Key Headings of an Outline

Title
1. Introduction and objectives
2. Methodology
3. Results and discussion
4. Conclusions
5. References

Step 2: Establish Minor Headings Within Major Headings

Arrange your minor points under their major headings. Often you will need more than two levels of headings, especially for complex subjects. It is normal to use Heading 3 and Heading 4 to better organize the writing in proper relationship. It is unusual for the headings to go beyond Heading 4; otherwise, the numbers become cumbersome.

Visuals are an integral part of your outline and should be put where they are expected to appear. Describe each visual with a tentative caption, e.g., "Schematic of...." All these are temporary and evolving like other information in the outline. You can add or remove visuals as needed.

At this step, make sure that each of your headings is marked with the appropriate sequential Arabic numbers combined with optional letters. You can cross-reference the decimal numbering system while you are developing the document. In addition, make sure that the headings have parallel structures.

In addition, this is the time to consider where to place the visuals (such as figures and tables) to achieve clarity in writing. You can simply reserve the space by writing *visual for* ..., noting the purpose of the figure, table, or another type of visual. When you begin drafting later, integrate the visuals smoothly into your body text.

Example 4.2 shows a sample short outline with three levels of headings. It has only Arabic numbers, but you can use letters if you prefer. You can use keywords to organize your rough ideas at the beginning. Later, you gradually add more phrases, clauses, and sentences and the ideas will become clearer and clearer too

Example 4.2. Short Outline with Headings, Keywords, and Visual Titles

Title: Cobalt-based solvents for absorption of nitric oxides
1. Introduction
1.1 Literature review
1.1.1 Motivation
    Air pollution, combustion, air cleaning...
1.1.2 Earlier experimental studies
    Adsorption, absorption, thermal conversion ...
1.1.3 Numerical works
1.2 Knowledge gaps
    Cost-effectiveness, technical feasibility...
1.3 Objectives
    Understanding mechanisms of...
1.4 Structure of the thesis

2. Methodology
2.1 Overall experimental setup
    Figure. Schematic of the test setup
2.2 Test apparatus
    Figure. Photo of the test apparatus
2.3 Instrumentation and data collection
    Table: List of devices and data collection
2.4 Data analyses and expected results
    Graphs here
    Tables here
3. Results and Discussion
[ensure to echo objectives]
4. Conclusions
5. References

Step 3: Create the Lengthy Outline with Sentences

What you have now is a short outline, and it can be extended into a lengthy outline with complete sentences and paragraphs without precise grammar, language, or punctuation; they will be taken care of when you revise the draft following the guidelines about paragraphs, sentences, phrases, words, and even punctuation. This conversion can be part of the drafting. However, a sentence outline is recommended because it may be the most challenging part of your writing. A complete lengthy sentence outline is almost like a *crude* draft (*see* Example 4.3). The topic or sentence outline should be flexible because it may change when you write the first draft, but all the first sentences tell a story.

There are various ways to prepare a final outline, and you are encouraged to create an outline following your style. Word processing software like *Microsoft Office Word* has a function that allows you to create and view the *Outline* of a document. Nonetheless, the organization of ideas should be clear and easy to follow at this point. Appendix A3 is a sample outline preceding its first draft.

Formatting the headings is optional, even though you might think it helps track their relative importance when you write the first draft. Formatting any part of the document at this stage is a waste of your time because you may change your mind while you are drafting. Headings are added, removed, or moved while you are working on the first draft. Therefore, you should format your document after finishing all revisions.

Example 4.3. Lengthy Outline with Topic Sentences

Title: Cobalt-based solvent for absorption of nitric oxides
1. Introduction
1.1 Literature Review
1.1.1 Motivation

- Air pollution impacts human health and the environment.
- Fossil fuel combustion is one of the sources of air emissions.

1.1.2   Earlier experimental studies

- There are several technologies available for air emission control. Selective catalytic reduction; wet scrubbing; low NOx combustion.
  (Visuals to show experimental setup)
- The power industry needs cost-effective technologies for nitric oxide (NOx) emission control.
- This work is focused on NOx emission control by wet scrubbing using cobalt-based solvents.
  (Table to compare the pros and cons of earlier works)

1.1.3   Numerical works
[*Repeat the same practice under other headings.*]

With established purpose, readers, scope, and outline, you are ready to draft your manuscript by expanding the outline from sentences to paragraphs. Meanwhile, you create and integrate visuals into the text. The next section introduces the practical techniques for first draft.

Note

A well-prepared outline should tell a whole engineering story. It clearly describes the reason for the research, used in the study, and key contributions to the field.

4.2   Outline-Based Drafting

An experienced writer begins drafting with the *body* of the document. Consider writing the *front matter* last when you have a clean draft. For example, many international graduate students start writing with the *Abstract* because they see the *Abstract* first in publications. On the contrary, you should draft the *Abstract* last. Nonetheless, the headings remain in the draft to divide the manuscript into sections. These sections may be short or long, depending on their functions in your document.

Your first draft follows the lengthy outline with sentences. Example 4.4 shows how two sentences in an outline are converted into two paragraphs, following the paragraph structures (*See* Chap. 11). Meanwhile, all sentences in the paragraphs should follow sentence structures (*See* Chap. 12).

To emphasize, you ought to ensure that each paragraph in your draft has a controlling idea supported by a body of evidence. To achieve clarity, convert each of the sentences in the outline into a paragraph with a single controlling idea. There are various techniques in constructing paragraphs. A typical paragraph comprises of *the topic sentence*, *the body sentences*, and *the concluding sentence*. The topic sentence is the statement of the controlling idea of the paragraph. The body sentences support the statement. The paragraph may end with a concluding sentence, which reiterates the main ideas of current paragraph or introduces next paragraph.

The first draft is a *rough* draft, which continues organizing your ideas. You can imagine that you are talking to your readers, who are sitting right in front of you. They would tolerate and understand even if you choose imprecise words in verbal communication. Therefore, you focus on the flow of ideas established in the outline, which serves as a road map for your presentation. At this point, you

> **Note**
>
> The first sentence of each paragraph remains as the corresponding sentence in the outline. You can certainly make reasonable revisions to the first sentences while drafting, but the controlling ideas should remain the same, unless you want to redo the outline.

Example 4.4. Converting Outline to Draft

The first two sentences in A3 Sample Outline are as follows:

- Lithium-ion battery (LIB) is an important energy storage and transmission technology in a modern society.
- The key elements in a LIB include a cathode, an anode, an electrolyte, and a separator.

These two sentences become two paragraphs in the draft:

Lithium-ion battery (LIB) is an important energy storage and transmission technology in a modern society. LIB has been widely used in portable devices, electrical vehicles, large-scale power sources, and the like because of its high energy density, long cycle life, and low cost.

The key elements in a LIB include a cathode, an anode, an electrolyte, and a separator. The cathode and anode receive and detach lithium ions during battery charging and discharging, respectively. The electrolyte soaked in the separator acts as the lithium ions reservoir, enabling the lithium ions to move between the two electrodes. The battery separator, located between the cathode and the anode, is a barrier that prevents the direct contact between the electrodes and short circuit of the battery. The separator is widely recognized as a crucial component related to the overall battery performances, including lifetime, safety, and energy density, although it does not directly participate in the battery reactions. (Li and Tan 2020)

---

pay little attention to the language details such as grammar, spelling, punctuation, format, and so on. Otherwise, you may deviate from the more important task, which is connecting ideas.

Language *is* important to your final publication, but language errors should not be your concern in the *first* draft. You start paying attention to language when you revise the draft manuscript. Ultimate refinements come with third or more revisions. Flawlessness can be reached after revisions, formatting, and proofreading.

4.3   Practice Problems

Question 1

Download the sample paper and follow these steps.

1. Create a short outline for the introduction section based on the keywords in the first sentences of the *Introduction.*
2. Create a long outline for the introduction section based on the first sentences. (Note: you do not need to use the full sentences.)

3. Read your outline and check the organization of ideas.
4. Closely examine the paragraphs in the sample paper, and check whether each paragraph has one single controlling idea as stated by the first sentence.

- **Sample paper:** Fuller SJ, Zhao Y, Cliff SS, Wexler AS, Kalberer M, 2012. Direct surface analysis of time-resolved aerosol impactor samples with ultrahigh-resolution mass spectrometry, Analytical Chemistry 84 (22): 9858–9864. DOI: https://doi.org/10.1021/ac3020615

Question 2

The following is an outline (the first sentences of paragraphs in draft) of the *Introduction* of a literature review article. Check the flow of thought (ideas organization) and see what is/are missing.

1. Introduction

- Cancer metastasis refers to that cells shed from primary tumor are circulated by blood system and invade other organs as new tumors.
- Enrichment of CTCs from patient blood is critical to accurate detection and characterization of CTCs.
- Figure 1 presents the classification of CTC enrichment methods; they are categorized into single- and hybrid-modality methods
  Figure 1 Classifications of CTC enrichment methods
- Despite substantial progresses in single-modality methods, most of them still face with limitations and have not been widely adopted in clinical utility.
- The hybrid-modality method with increasing popularity can yield better enrichment performance [32–34].
- Many review articles for CTC enrichment are available in the literature. … However, there are limited reviews on hybrid-modality methods.
- The rest of this article is organized as follows.

Drafting Introduction

5

5.1   Sources of Information

Introduction or *Literature Review* can help better understand the state of the art. The background information collected and reviewed should be from various sources, including online and printed documents, for comprehensiveness. Some researchers may also consult internal archives for the most relevant information. While peer-reviewed articles and books are credible sources of literature, peer-reviewed documents published on academic and government websites are also acceptable. Furthermore, libraries are recommended for locations where online resources are limited.

> **Note**
>
> Many published works contain errors or outdated information or both. You must cross-examine the key information you will use because sciences and technologies advance every day.

You must ensure the validity of sources. In some countries, online resources are abundant and easy to access, but it is not always the case in other countries. The internet provides access to most information that we need. However, it is challenging to verify the completeness and accuracy of some information because anyone can publish online. A safe approach to valid information is to access credible sources. The online versions of articles in reputable journals, books by prominent researchers, and so on would have the same merits as the printed version.

Regardless of the scope, keep in mind that the more state-of-the-art the information, the better. Using outdated information generally undervalues the value of your work because there is a high probability that you are missing recent advances in the same field of research. Lacking state-of-the-art information may also be viewed as incapability or dishonesty of the author(s). To emphasize, writing reveals your capability, personality, and professionalism.

Introduction varies in length, depending on its purpose. It can be one paragraph in a cover letter, one section in a short document, or one chapter in a thesis or book. The *Introduction* of a short document may be merged with *Literature Review* in journal articles. *Literature review* may also become a stand-alone article for the synthesis of knowledge, but it is beyond the focus of this book.

© The Author(s), under exclusive license to Springer Nature Switzerland AG 2022
Z. Tan, *Academic Writing for Engineering Publications*,
https://doi.org/10.1007/978-3-030-99364-1_5

5.2   Funnel Approach to Literature Review

A comprehensive literature review, as the core of the *Introduction*, justifies the importance and highlights the novelty of your work. Literature review is more than a compilation of earlier publications. As you review each source, evaluate its validity, investigate its limitations, and synthesize its value to the readers. You can chronologically arrange your evaluation or subcategorize them into different topics.

Figure 5.1 illustrates the *funnel approach*, which is an important technique for drafting the *Introduction* and *Literature Review*. A thorough *Introduction* begins with a relatively big picture and importance to the field, gradually narrows to the state of the art and their limitations, funnels naturally to knowledge gaps, and ends with focused objectives. At the end, you articulate the questions to be answered, the problems to be solved, the technologies to be developed, the methodology to be improved, and so on. Finally, a transition naturally leads your writing to the next part of the document, *Objectives*.

The funnel approach is critical to the justification of your research and the publication of your work. It is also important to the clarity of your writing. Thus, this technique is further explained with Examples 5.1, 5.2, 5.3, 5.4, 5.5, 5.6, 5.7, and 5.8 (Source: Fuller et al. 2012).

Example 5.1. The First Paragraph of the *Introduction*
Aerosol particles are <u>important</u> components of the <u>atmosphere</u> and strongly influence the Earth's <u>climate</u> by directly scattering or absorbing sunlight and indirectly by aerosol–cloud interactions. Aerosol particles are <u>also</u> involved in negative <u>health</u> effects caused by air pollution and are linked to increases in respiratory and cardiovascular diseases. All these particle effects are influenced by the <u>chemical composition</u> of the <u>aerosol particles</u>. A major fraction, often more than 50% in mass, of tropospheric aerosol is <u>organic material</u>, <u>which is very poorly understood</u> on a molecular level, although several thousand compounds have been separated with chromatographic techniques. To identify sources of particles, but also to understand compositional changes during their atmospheric lifetime, an automated method with high time resolution is <u>highly desirable</u>.

Comment:
This *Introduction* begins with the *importance* of the work (*atmosphere, climate, health,* etc.). Then, it narrows it down to *chemical composition of aerosol particles* and highlights the limitations in the field (*organic materials, poorly understood*). At the end, they bring up the need of the work (*automated method, high resolution, highly desirable*).

Fig. 5.1 Funnel approach to *Literature Review*

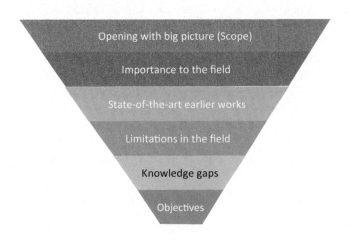

The first one to two paragraphs to follow usually are a high-level description of the background information. However, this prominent journal does not give the authors much room for the elaboration on the big picture. Most writing requires elaboration on the "big picture." Thus, each of the sentences in this paragraph could be extended into a paragraph of a long document such as a thesis or a book.

Example 5.2. The Second Paragraph of the *Introduction*
The analysis of the organic fraction of atmospheric aerosol particles on a molecular level is often <u>challenging</u> because their <u>composition</u> can be highly variable in time and space and usually only small sample amounts are available for analysis because atmospheric concentrations are typically a few micrograms per cubic meter. <u>The composition also depends on the size fraction</u> of the aerosol; for example, resuspended, wind-blown particles are mostly larger than 1 μm and have often a very distinct chemical composition compared to particles smaller than 1 μm, which are mainly emitted by combustion sources or formed by chemical reactions in the atmosphere.

Comment:
Following the first paragraph, the ideas in the second paragraph become more specific by identifying the *challenges* and the general *scope* of work.

Example 5.3. The Third Paragraph of the *Introduction*
Thus, analytical–chemical techniques used to analyze atmospheric <u>organic aerosol particles need</u> to be <u>highly sensitive</u> to allow for highly time-resolved analyses at trace concentrations. In addition, the analysis technique needs to be able to characterize highly complex organic compound mixtures with thousands of mostly unknown components.

Comment:
This paragraph narrows down its controlling idea from *aerosol is important* to *highly sensitive analysis of organic aerosol particles is needed*. The authors seem to set the stage for the major contributions to be introduced in *Conclusions*. However, this short paragraph could be better structured if it follows typical paragraph structure.

Example 5.4. The Fourth Paragraph of the *Introduction*
<u>There are a number of analytical–chemical techniques</u> that allow measuring the composition of aerosol particles with high time resolution. <u>Online aerosol mass spectrometry</u> (AMS) techniques, for example, allow for very highly time-resolved particle composition studies. <u>However,</u> such measurements are demanding with respect to manpower and other resources and are usually performed only for a few weeks at a specific site. Thus, such measurements rarely provide insight into long-term trends of particle composition. <u>In contrast, aerosol samples collected on filters or impactors</u> over long time periods are more readily available <u>but</u> have mostly a rather low time resolution of the order of a day or more, and their chemical analysis usually involves time-consuming.

Comment:
This paragraph summarizes earlier works closely related to the scope of the current study. The authors provide criticism (introduced by *however*, *but*) and evaluate the limitations of the state of the art. They imply the novelty of the current study.

Nonetheless, a literature review is more than a simple compilation of earlier works. It is essential to show your *critical thinking* instead of accepting what you read at the surface. A researcher is a critical thinker. Thus, you must comment on the state of the art and identify the limitations, which are also new research opportunities. The limitations are closely related to your own study, which addresses the knowledge gaps between earlier studies and yours. Example 5.5 (same text as Example 5.4) illustrates how to identify the knowledge gaps and the research opportunities in the field of study.

Example 5.5. Identifying Knowledge Gaps

[This is the same paragraph as in Example 5.4] There are a number of analytical–chemical techniques that allow measuring the composition of aerosol particles with high time resolution. Online aerosol mass spectrometry (AMS) techniques, for example, allow for very highly time-resolved particle composition studies. However, such measurements are demanding with respect to manpower and other resources and are usually performed only for a few weeks at a specific site. Thus, such measurements rarely provide insight into long- term trends of particle composition. In contrast, aerosol samples collected on filters or impactors over long time periods are more readily available but have mostly a rather low time resolution of the order of a day or more, and their chemical analysis usually involves time-consuming workup and is prone to artifacts during sample preparation.

The underlined sentences in Example 5.5 identify the limitations of the earlier works and prepare the context for knowledge gaps to come. The limitations of earlier studies are usually introduced by conjunctions like *however*, *but*, *on the other hand*, *nonetheless*, and so on. (*See* Table 12–1. Conjunctions). However, you should avoid introducing limitations using the word *unfortunately* or similar negative phrasing, although it appears in many published works. (*See* 10.6. Positive and Negative Tones.)

Example 5.6. Avoid *unfortunately*

Emotional: Unfortunately a great variation of these parameters can be found in literature, and it is therefore difficult to compare the activities of catalysts tested in different laboratories. (Fino et al. 2016)

Correct: However, a great variation of these parameters can be found in literature. Therefore, comparing the activities of catalysts tested in different laboratories is challenging.

Example 5.7. Reviewing Closely Related Earlier Works

[Fifth paragraph] The rotating drum impactor (RDI) is a sampling technique that allows for off-line chemical analysis of particles with a time resolution on the hour time scale, which is sufficient to resolve most atmospheric aerosol processes. The RDI has been combined with synchrotron X-ray fluorescence (s-XRF) analysis to measure changes in metal content directly from the RDI samples. Only one study analyzed so far organic components collected with an

RDI: Emmenegger et al. investigated the temporal variability of polycyclic aromatic hydrocarbons collected at an urban location with two-step laser mass spectrometry, where an IR laser was used to desorb analytes directly from the RDI stripes without further sample preparation.

[Sixth paragraph] <u>The recent development of commercially available surface mass spectrometry ionization techniques</u> such as liquid extraction surface analysis (LESA) and desorption electrospray ionization (DESI) enables analysis of a wide range of organic compounds with high spatial and therefore high time resolution from RDI samples (for details see the Experimental Section). <u>LESA has been used previously for</u> the analysis of biological samples and pesticides, but applications for environmental samples have <u>not yet</u> been described in the literature. <u>Another online extraction technique, nano-DESI, was recently presented by Roach et al.</u> and was applied to atmospheric aerosol filter samples. Importantly, LESA and (nano-) DESI require no off-line sample preparation, such as solvent extraction and solvent evaporation. Reducing the number of sample preparation steps also reduces the possibility of introducing artifacts.

In engineering studies, there must be other researchers who attempted to address similar knowledge gaps. Therefore, following the fourth paragraph, the fifth and the sixth paragraphs in the sample paper (Fuller et al. 2012) identify and evaluate the most recent and the most relevant earlier works that seem to address the knowledge gaps. This step precedes the declaration of *Objectives*.

5.3   Drafting Objectives

The state-of-the-art literature review eventually funnels down to the objectives. The objectives must be *specific* to *the* work; they must be focused and closely related to the scope of your own study as indicated by the title. You can choose some scientific questions to answer through your work for the coherence of writing. You should separate them into different short documents (articles) or different sections (chapters) of a long document if you have multiple irrelevant objectives in mind.

The introduction section normally ends with objectives – avoid a sudden stop in your literature review. Example 5.8 shows how Fuller et al. (2012) funnel their preceding review down to their objectives.

Example 5.8. Funneling Down to the Objectives
<u>For this study</u> RDI collection was combined <u>for the first time</u> with LESA and ultrahigh-resolution mass spectrometry (UHRMS) to <u>characterize organic components in aerosol particles and to observe changes in ambient concentration with a time resolution of about 2.5 h for sub- and super-micrometer particles.</u> UHR-MS allows identification the elemental composition of unknown compounds and is a valuable technique for the analysis of highly complex and often poorly characterized atmospheric organic aerosols where often the majority of <u>the compounds are unknown</u>.

Comment:
The phrase "For this study" declares that the objectives are specific to this publication. You can also use *in this paper, in this work, in this thesis*, etc. I see two objectives in this paragraph. The first one is to develop a new technique (… *for the first time*), and the second, to characterize *the organic compounds in aerosol particles….*

5.4   Errors in Introduction and Objectives

Examples 5.1, 5.2, 5.3, 5.4, 5.5, 5.6, 5.7, and 5.8 illustrate the well-written *Introduction* and how it funnels down to the *Objectives*. This *Introduction* section is well-written because most of the co-authors are native English speakers and well-established researchers in their field. However, it is more challenging than it appears to many writers.

All writers make mistakes in writing, and you will make fewer and fewer mistakes with continual learning and practice. New and intermediate-level writers often make the following mistakes writing *Introduction* and *Objectives,* especially typical to non-native English speakers:

1. **Unclear funnel shape**

The *funnel approach* is critical to the effectiveness of the introduction, although this structure is not always necessary to writing in another culture or language. A broken funnel may result from one or more of the following errors:

- Insufficient breadth of background, at the top of the funnel
- Insufficient explanation of importance, in the middle of the funnel
- Insufficient justification of the novelty, near the bottom of the funnel
- Insufficient summary of prior arts
- Unclear knowledge gaps
- Multiple directions leading to irrelevant questions to be answered
- Incomplete literature review without clear goals, when the literature review stops suddenly

2. **Overly general objectives without focus**

Example 5.9 shows a typical mistake in overly general objectives.

Example 5.9. Overly General Objectives
......~~Unfortunately~~, by digging out in literature, we found that a detailed examination to the effect of drag models on CFD modeling seems unavailable. In spouted bed systems, the volume fraction of particles can vary from almost zero to the maximum packing limit, leading to much more complex behavior of drag forces than that in normal fluidization systems. By incorporating various drag models into the two-fluid model, the present study is conducted with the <u>aim of fully understanding</u> the influence of the choice of drag models on simulation and thereby laying a basis for the CFD modeling of spouted beds. (Du et al. 2006; reprinted with permission)

Comments:
It takes numerous studies to understand part of a discovery. Similarly, the basis for CFD modeling should have been established by other researchers too. The specific incremental contributions in one study should become the specific objective(s).

Many researchers have developed or are developing novel CFD models for a variety of applications. It is unlikely for one to fully address the complexity of this problem. You need to be specific on the questions to be answered or the problem to be solved by *your* work being presented.

Example 5.10. Avoid Overly Broad Objective
Incorrect: The objective of this research is to develop <u>a novel CFD model</u>.
Correct: The objective of this thesis research is to develop <u>a drift- flux model for indoor aerosol dispersion in classrooms</u>.

---

3. **Confusing Tasks with Objectives**

New and intermediate writers often confuse the tasks with the objectives. A task describes how to study, not why to study. The task is part of methodology, indicating the investigations needed to achieve a specific objective. For example, Example 5.11 compares the task-like objectives listed in a draft to those that appear in the final version. Words like *develop*, *measure*, and *analyze* normally indicate tasks. Conversely, words like *evaluate*, *understand*, and *investigate* often indicate the actual objectives behind the tasks.

Example 5.11. Tasks vs. Objectives

Tasks	Objectives
Although many studies have been conducted to convert biomass to biooil, no data is available in literature about the hydrothermal conversion of cattle manure. This research will focus on cattle manure with the following three specific objectives.	Although many studies have been conducted to convert biomass to biooil, no data are available in literature about the hydrothermal conversion of cattle manure. This research focuses on cattle manure with the following three specific objectives.
(1) To develop a 1.8-L batch reactor for HTC of cattle manure.	(1) To evaluate the feasibility of HTC of cattle manure (using a 1.8-L batch reactor).
(2) To measure the HTC conversion rate and purity using the 1.8 L reactor	(2) To evaluate the economics of HTC of cattle manure (based on data obtained using the 1.8-L batch reactor).
(3) Analyze the chemical composition of the HTC products using a tubing reactor under different conditions	(3) To investigate the mechanisms and reaction pathway of HTC of cattle manure (using a 76 mL tubing reactor).

Note

Sometimes a novel model, procedure, apparatus, or the like may be the objective of a study that enables the works to follow. For example, the objective of a master of science thesis posted at http://hdl.handle.net/2142/16828 aims to build an experimental setup and develop a procedure that can be used by other researchers in the same field. It is part of a larger research project; the setup presented in the thesis was used later in some follow-up studies. (Source: personal communication)

5.5 Practice Problems

Question 1 Identify the *scope, importance, state of the art, limitations, knowledge gaps,* and *objectives,* if any, in these two sample papers:

- **Sample paper 1:** R. K. Srivastava & W. Jozewicz (2001) Flue gas desulfurization: the state of the art, J of A&WMA 51: 1676–1688. https://doi.org/10.1080/10473289.2001.10464387
- **Sample paper 2:** H-H Tseng, M-Y Wey, C-L Lin, Y-C Chang (2002) Pore structure effects on Ca-based sorbent sulfation capacity at medium temperatures: activated carbon as sorbent/catalyst support, J of A&WMA, 52: 1281–1287. https://doi.org/10.1080/10473289.2002.10470859

Question 2 Highlight the *objectives,* if clearly stated, of these two papers: Following the guide in this book, match the statement of objectives written in these papers with these errors:

1. Missing objectives
2. Overly general
3. Mistaken tasks with objectives
4. Disconnection between knowledge gap and objectives

- **Sample paper 1:** R Imhoff, R Tanner, R Valente, M Luria, 2000. The evolution of particles in the plume from a large coal-fired boiler with flue gas desulfurization, J. of A&WMA, 50: 1207–1214. https://doi.org/10.1080/10473289.2000.10464153
- **Sample pager 2:** S Kiil, M Michelsen, K Dam-Johansen, 1998, Experimental investigation and modeling of a wet flue gas desulfurization pilot plant, *Ind & Eng Chemistry Res* 37: 2792–2806. https://pubs.acs.org/doi/full/10.1021/ie9709446

Question 3 Access the sample paper at this website: tandfonline.com/doi/pdf/10.1080/027868201753306750

1. Extract the first sentence of each paragraph in *Introduction*.
2. Construct a new paragraph by putting these first sentences in the order of appearance.
3. Read the new paragraph you constructed and feel the organization of ideas.
4. Identify the ideas in the paragraphs with the funnel shape: categorize them into *scope, needs, state of the art, critics, knowledge gaps,* and *objective*s.
5. Following the guide in this book, identify and revise the writing errors in the last paragraph of *Introduction*.

- **Sample paper:** D Kane, B Oktem, M Johnston, 2001. An electrostatic lens for focusing charged particles in a mass spectrometer, *Aerosol Science &Tech* 35 (6): 990–997

Question 4   Access the sample paper at this website: pubs.acs.org/doi/abs/10.1021/es001323y

1. Extract the first sentence of each paragraph in *Introduction*.
2. Construct a new paragraph by putting these first sentences in the order of the appearance.
3. Read the new paragraph and check the organization of ideas.
4. Identify the ideas in the paragraphs with the funnel shape: categorize them into scope, needs, state of the art, critiques, knowledge gaps, and objective.
5. Following the guide in this book, identify and revise the writing errors in the last paragraph of *Introduction* section.

• **Sample paper:** D. Kane, M Johnston, 2000. Size and composition biases on the detection of individual ultrafine particles by aerosol mass spectrometry, *Environmental Sci & Tech* 34: 4887–4893. DOI: 10.1021/es001323y

Question 5   Convert the tasks into objectives by rephrasing Example 5.11.

Drafting Methodology

6

6.1 Research Methods

Numerous research methods are available for engineering researchers, including analytical, numerical, and experimental approaches. Regardless of the methods used, clear description of methodology is important to training new researchers, sharing new knowledge, and transferring novel technology that benefit humankind. This applies to most original engineering publications, including journal articles, graduate theses, technical books, etc.

Regardless of the method, *repeatability, replicability,* and *reproducibility* are essential to engineering research. The readers of engineering publications are more than the experts. They are aimed at diverse readers, including undergraduate and graduate students. Your methodology trains these new researchers entering your fields of research. Clarity even becomes an ethics matter for description of experimental procedures that involve safety risks.

The methodology for experimental studies must contain sufficient details, allowing the readers to replicate the data and validate the results if they wish to do so. Otherwise, the results are considered questionable. The details include equipment and devices, materials and supplies, and the apparatus and components used in the experiment.

In addition, you need to justify your choices of approach to the research by comparing with alternatives reported in earlier publications. You also should explain the novelty of your approach. All these factors matter to the accuracy of your results and the validity of your conclusions. In addition, you should include the observations and problems encountered, if any. They save your readers' time from repeating the same mistakes in their studies.

The principles described in preceding paragraphs are applicable to all research methods. However, this book excludes analytical methods because of its less-frequent stand-alone use in engineering studies. You can read books focused on research methods such as the one by Thiel (2014) for systematic training.

6.2 Describing Numerical Approach

Both numerical and experimental methods are widely used in modern engineering research. For description of numerical models, you should clearly explain the physical models and simplifications made to solve the equations. In short, you need to clearly document the following information:

Z. Tan, *Academic Writing for Engineering Publications*,
https://doi.org/10.1007/978-3-030-99364-1_6

1. **The approach developed and used in your study**

Typical numerical models are developed from theoretical analyses using many equations. These equations describe the general principles in the fields of research, such as conservations of mass, momentum, and energy. They are usually simplified to match the conditions of the specific problems under investigation.

2. **Assumptions and simplifications for numerical models**

General principles cannot solve a specific problem without assumptions or simplifications or both. Most likely you need to make reasonable assumptions and simplifications before you can proceed with a numerical simulation. Various factors may affect the accuracy of your results and conclusions, but simplifications make your studies affordable and acceptable. It is a compromise between accuracy and cost. Therefore, you must clearly describe and justify the assumptions and simplifications. In addition, you need to detail the boundary conditions and initial conditions. Different simplifications and conditions may lead to different results and conclusions.

3. **Platform used to solve the equations**

You may use the latest commercial software to solve your simplified equations; clearly write down this information. If you import user-defined functions or code by yourself, write this detailed information down too. These details are important to the clarity of your writing, the accuracy of your results, and the applicability of your conclusions. Readers rely on these details to judge the quality of your work, to replicate the results, and to identify the limitations in your research. (*See* Chap. 5.)

4. **Model validation**

"How much can I trust the model results?" Your readers may ask this question in their mind when they read your document. In the scientific and engineering world, seeing is believing.

Model validation is important and challenging to some numerical studies. Nonetheless, you must validate your models first before using them as a tool for further investigations. If not by your own experimental data, you can validate your models with other models or experimental data in the literature. Data do matter. Otherwise, it is viewed as overconfidence if you draw a conclusion without evidence that your results are trustworthy.

In addition, you need to explain the error sources and their relative importance to the accuracy of your results. They may include the simplifications, assumptions, and so forth. You ought to point them out instead of leaving them to the readers to guess.

Many well-written articles based on numerical research are available in the literature. To start, the paper by Zhang et al. (2020) can help you understand how to describe the methodology of a numerical work (dx.doi.org/10.1177/1420326X20941166).

6.3   Replicability

Replicable results are essential to engineering research. Therefore, a well-written methodology contains precise descriptions of experimental procedures. Specifically, the writing is centered on three *Ws*: *where*, *when,* and *how* to conduct the experiment. This information also helps the readers analyze the accuracy of results.

As introduced in *Numerical Approach*, the description of experimental approach should also have sufficient details allowing the readers to repeat procedure if they decide to do so. Thus, treat your *Methodology* section as a user manual; any reader can repeat your data collection by following the "manual."

A clear description of *Methodology* normally details the following key information. Lacking one or more of the entries on the list reduces the clarity of writing and the replicability of the results:

1. *When* and *Where,* if important to result
2. Description of experimental setup, if any, from overall to its components
3. Function of key components
4. Test conditions and justification of parameters
5. Acceptance criteria for conditions (e.g., *accurate, stable, equilibrium*)
6. Models, suppliers, and key specifications of devices
7. Justification of the choice of the instruments
8. Data analysis tools and methods
9. Expected results, if possible
10. Replicates of dataset (minimum three replicates), if applicable
11. Key sources of errors and uncertainties of the results
12. Reasons for excluding certain results, if applicable
13. Preliminary trials to minimize errors, if applicable

The *Methodology* section in the sample paper by Fuller et al. (2012) has most of this key information. Example 6.1 matches the key information in the preceding list with the text in the *Methodology*. It is a well-written *Methodology* section because of the level of detail in the test procedure and data analysis.

The preceding example shows the level of detail in a well-structured *Methodology* section. As seen on the right side of the text, the underlined text details and justifies the methods used for data collection and analysis. Some details might look trivial and unnecessary to you, but well-trained researchers understand that something trivial might cause big errors. Thus, it is better to include these details, allowing your readers to evaluate the quality of your work. To further ensure clarity, the authors use *Supplemental Information* for nonessential information, which might be useful to other researchers.

Nonetheless, this writing still has very minor errors, although the authors are native English speakers. First, they use a subjective word *unfortunately*. Second, a list would have enhanced the ease of comprehension for the description of the eight-stage *RDI sampler* and *UHR-MS Data Analysis*. With the page limit set by the publisher, the authors chose sufficient information (clarity) over reading experience, I believe.

6.4   Precise Description of Method

Precise description is crucial to the engineering writing aimed at readers who are unfamiliar with the research. Treat your readers as if they were first year college students. You should avoid the assumptions that your readers know the same background information, are capable of data analysis, or can draw correct conclusions from your data.

Specifics are important to replicability of the work. They also allow the readers to evaluate the accuracy of results, the validity of argument, the need of further study, and so on. On the contrary, overly general description lacking specifics is a typical mistake that non-native English writers may make.

Example 6.1. Well-Written *Methodology*

Original writing	Analysis
Sampling Site Location and Sample Collection. RDI aerosol samples were collected at a site on the <u>southern border of a dairy farm</u> with about <u>1200 milking cows</u> in the San Joaquin valley in California, one of the most intensively farmed regions in the United States. <u>The sampling site is discussed in more detail in the Supporting Information and in Zhao et al. (2012)</u>	*Where* *The number of cows may affect the results.* *SI and cross-reference for clarity*
An RDI, based on a design by Lundgren was used to collect size-segregated aerosol samples for <u>approximately 2 weeks from 11 a.m. 5/25/2011 to 11 a.m. 6/7/2011</u>. The fourth-generation RDI used in this study <u>is described in detail in Zhao et al. (2012), and only a short description is given here</u>. Rather than collecting all particles on the same impaction spot during the sampling time, a Mylar strip on a slowly rotating drum is used as the impaction surface resulting in a time-resolved deposition of the impacted particles. Ambient air is sampled at <u>a flow rate of 16.7 lpm</u>, and aerosol particles are collected via impaction <u>on greased (Apiezon L high-vacuum grease) Mylar strips</u>. The Mylar is greased <u>to increase the collection efficiency by reducing particle bouncing. The time resolution is determined by the rotation speed of the RDI drums combined with the aerodynamic spread of sample deposition. This varies with the particle size cut, with smaller particle sizes giving higher resolutions.</u> The RDI sampler consists of eight stages collecting samples in eight size ranges (<u>i.e., stage 1, 10–5 μm; stage 2, 5–2.5 μm; stage 3, 2.5–1.15 μm; stage 4, 1.15–0.75 μm; stage 5, 0.75–0.56 μm; stage 6, 0.56–0.34 μm; stage 7, 0.34–0.26 μm; stage 8, 0.26–0.09 μm</u>). Across all eight stages the type of deposition could be divided visually into two distinct sample types, first on the stages collecting particles with larger diameters (stages 1–5), brown colored particles follow a similar deposition variability with time. On stages with smaller diameter particles (stages 6–8) particle deposits were much darker and also showed a time variation similar to each other that was distinct from that observed on stages 1–5. <u>Stage 3 and stage 8, corresponding to the size ranges of 2.5–1.15 and 0.26–0.09 um, respectively, were chosen as representative of the two deposition types and were investigated further in this study</u>.	*When* *Cross-ref to an earlier publication.* *How* *Details* *Model and supplier* *Justifying choice of method* *Justify choice of device; explain probable source of error.* *Detailed specifications of device* *Preliminary trail to justify choice of parameters* *Justify choice of parameters and the sampling methods.*
Sample Analysis. The organic composition of particles deposited on stages 3 and 8 of the RDI was analyzed with mass spectrometry using a <u>new direct extraction technique</u>, LESA. LESA is a mode of operation of the TriVersa NanoMate chip based nano electrospray ionization (ESI) source <u>(Advion, Ithaca, USA)</u>. Samples are mounted onto a movable sample stage, and a <u>small amount of solvent (1–10 μL)</u> is then dispensed from a pipet tip on the surface of the sample <u>(Fig. 1)</u>. The micro liquid junction is maintained to extract analytes present at (or close to) the surface of the sample and to dissolve them into the small solvent extraction volume. The droplet is aspirated and subsequently sprayed via an infusion method utilizing the chip-based ESI22 ... All mass spectra shown here were recorded in positive ionization mode.	*Emphasis on <u>new</u> technique for sample analysis* *Model; supplier* *Specific values* *Cross-ref. (Fig. 1) for clarity.*

A section of each Mylar RDI sample strip was cut to <u>size (20 mm × 70 mm) and mounted on a standard glass microscope slide before being analyzed by LESA at 20 points along a linear path at 1 mm intervals, which corresponds to the approximate liquid junction droplet size as shown in Fig. 2.</u> This corresponds to a sampling <u>start time of about 19:30 on May 26, 2011 and an end time of 21:50, May 28, 2011, with a sampling resolution of approximately 2 h 40 min.</u> Blank extractions were taken from an area at the edge of the strip where no sample was deposited. Extraction volumes of <u>1 μL of solvent (80:20 methanol/water)</u> were deposited at a height of 0.8 mm from the sample surface to form the liquid junction which was maintained for <u>30 s. This time allowed for analytes present at or close to the surface of the RDI strips to dissolve. Only very limited loss of solvent and resolution was observed due to droplet spreading. Many previous studies using LESA have repeatedly deposited and aspirated solvent onto the sample on a single extraction spot to aid mixing of the extracted sample into the solvent within a short contact time of typically 1–5 s. However, this leads to sample loss through each deposition/aspiration cycle as a small amount of solvent is lost to the surface each time the sample is aspirated. However, a single deposition and aspiration reduces sample loss while increasing the contact time with the surface allows a longer period for analytes to dissolve and sufficient time for sample mixing through diffusion due to the small extraction volume. Contact times of over 30 s are less effective due to breakdown of the liquid junction.</u>	*Detail sample method.* *Detail time, which may affect accuracy. Resolution may affect accuracy.* *Details the method* *Justify choice of method with trial tests and similar methods in the literature. Part of error analysis*
An ultrahigh-resolution mass spectrometer (<u>LTQ Velos Orbitrap, Thermo Scientific, Bremen, Germany</u>) with a <u>resolution of 100000 at m/z 400</u> and a typical mass <u>accuracy of ±2 ppm</u> was used to analyze the organic compounds present in the samples following extraction by LESA. <u>The resolution and mass accuracy of UHR-MS allows the identification of the elemental composition of unknown organic compounds.</u> Samples were sprayed at <u>a gas (N2) pressure of 0.30 psi at 1.8 kV</u> in positive mode using a NanoMate nano-ESI source.	*Model and supplier information of device* *Resolution that affects result accuracy* *Justifying choice of equipment* *Details gas and condition*
UHR-MS Data Analysis. For each extraction point on the RDI stripes mass spectra were recorded for a <u>1 min infusion duration in a mass range of m/z 50–500. Almost no peaks above m/z 350 were detected, and therefore only peaks below m/z 350 are discussed in the following. This mass range is a frequently observed characteristic of ambient aerosol composition.</u> The instrument was calibrated to within <u>±2 ppm using a standard calibration solution as prescribed by the manufacturer. Molecular formula were assigned within a ± 2 ppm error and within the following restrictions: number of carbon-12 atoms = 1–20; carbon-13 = 0–1; sulfur = 0–1; sodium = 0–1; and the following elemental ratios of H/C = 0.2–3, O/C = 0–3, N/C = 0–1. Peaks and assignments not following the nitrogen rule or containing C-13 were not considered further.</u> In addition, peaks where no formula could be assigned within the restrictions mentioned above were also removed. Due to the low mass range of the detected peaks (below m/z 350) and the high accuracy of the instrument, multiple assignments are rare after considering the restrictions listed above. <u>When several formulas satisfied all restrictions within 2 ppm, then the formula with the lowest mass error was assumed to be correct.</u> Unfortunately, <u>due to the generally low peak intensities of the identified species, MS/MS analysis for further structural identification was not possible. Only about 10–15% of the peaks contain a sulfur atom and are not further discussed here. See the Supplemental Information for 46 more details on the data analysis procedure.</u>	*Data analysis* *Details time* *Justify the method used for analysis by preliminary tests and similar method used in the literature.* *Calibration to ensure accuracy and precision* *Justify choice and exclusion of data.* *Criteria of conditions for validity* *Assumptions or uncertainty* *SI to enhance clarity*

The key to precise description of a procedure, for example, is the accurate presentation of details. The process description begins with a system and function followed by those components. The descriptions can be one paragraph for a simple system or multiple paragraphs for a complex one, but all descriptions should integrate each component into the function of the system.

Example 6.2 compares vague with clear descriptions. More examples for writing with sufficient detail are available throughout this book. Some may be presented with emphasis. You should be able to master this type of writing skills with continual practice. (*See* 12.7.2)

Example 6.2. Vague vs. Precise Descriptions

Vague description	Precise description
A tall spray dryer	A 10-meter-tall spray dryer
Plant	Power plant
Unit	Heat exchanger
Unfavorable weather conditions	Thunderstorm
Structural degradation	A leaky pipe
Good performance	95% efficiency

In contrast with Example 6.1, Example 6.3 shows a poorly written *Methodology* section (Ge et al. 2015). Besides the language errors, this *Methodology* section lacks replicability because the writing is vague, qualitative, and overly general. For example, the authors should have included this information to enable data replication:

Models, suppliers, and specifications of the materials, devices, and instruments. For example, the readers may ask these questions: Is the *air* used pure air or room air? What kinds of *water bath*, *chamber*, and *flow meter* are used, and where to buy them with the same function?

Example 6.3. *Methodology* **Lacking Details**

Original writing	Details needed
2.2. Experimental method Air was pumped and divided into two routes. One route of air was pumped through a water bath tower containing benzene solution to take gaseous benzene into a mixing chamber, and then the completely mixed gas was introduced into reaction chamber. The concentration of benzene in reaction chamber could be adjusted by a mass flow controller. Another route of air directly passed through a water bath container without benzene solution for carrying water vapor. The RH in the reaction chamber could be controlled from 20% to 60% by another mass flow controller. The air in the reaction chamber was circulated by a flow fan facing discharge gap to keep uniform relative humidity and stable discharge condition. The gas flow rate in the discharge channel was 2.4 m/s, and the temperature of gas was 20 ± 1 °C. After adding the catalyst to the system, the reaction was started by emerging the DC power supply. When the concentration of benzene in the reaction chamber reached a steady state, the benzene over the catalyst and organic glass chamber surfaces reached the adsorption-desorption equilibrium.	*Details of air, pump, water bath, etc.?* *Concentration of benzene solution?* *Criteria for complete mixing?* *Details of the mass flow controller?* *Why 20–60%? How to measure?* *What type of flow fan?* *Why not 2.5? How to measure?* *What catalyst?* *Details of power supply?* *Criteria of steady state and equilibrium?*

Justification of chosen parameters. For example, why *20–60%* of RH? What would happen if RH were 80% or 10%? Why *2.4 m/s* instead of 2.5 m/s?

Criteria of conditions. How can a reader determine whether the system reaches *equilibrium* or *complete mixing*?

6.5   Practice Problems

Question 1 Identify the sentences, if any, that highlight the following techniques used in writing *Methodology* in sample papers below or any similar writing you can find:

1. Supplier information and key specifications of the devices and instruments used
2. Justification of choice of parameters
3. Criteria of conditions

- **Sample paper #1:** R. Imhoff, R. Tanner, R. Valente, M. Luria (2000) The evolution of particles in the plume from a large coal-fired boiler with flue gas desulfurization, J of A&WMA, 50:7, 1207–1214. https://doi.org/10.1080/10473289.2000.10464153
- **Sample paper #2:** M. Luria, R. Imhoff, R. Valente, W. Parkhurst, R. Tanner (2001) Rates of conversion of sulfur dioxide to sulfate in a scrubbed power plant plume, J of A&WMA, 51:10, 1408–1413. https://doi.org/10.1080/10473289.2001.10464368

Question 2 Examine Sample paper #1 in Question 1, and comment on whether it contains sufficient details allowing readers to repeat the experiment. If not, give three examples that are needed.

Question 3 Examine the *Methodology* of Sample paper #3, and identify the following writing mistakes following the guide in this book:

1. Lacking details in the experimental procedure
2. Qualitative instead of quantitative information
3. Omission of suppliers of materials, devices, and instruments
4. Missing justification of chosen parameters

- **Sample paper #3:** Ge H, Hu D, Li X, Tian Y, Chen Z, Zhu Y. 2015. Removal of low- concentration benzene in indoor air with plasma-MnO_2 catalysis system, *Journal of Electrostatics* 76: 216–221. https://doi.org/10.1016/j.elstat.2015.06.003

7.1 Data Validation

Invalid data may lead to inaccurate or incorrect conclusions. Thus, you must exclude invalid data from further analysis. For example, Fig. 7.1 shows the curves of three identical membrane samples tested by an external laboratory. The stress-strain curves for Samples 1 and 2 are similar, and they look reasonable by preliminary analysis. However, the curve for Sample 3 is different. To validate the data, we contacted the external lab for detailed test procedure and visual observation in the test. Then we learned that data for Sample 3 are invalid because of inconsistent handling of the samples. In this case, we must exclude data for Sample 3 from further study.

Nonetheless, the data validation process should be clearly documented for future reference. If you are writing a thesis, you should keep the invalid curve for Sample 3 in the same figure and explain why in the text. However, you may omit this step for a journal article because of page limit. This omission concentrates your attention on the objectives of your paper.

7.2 Aiming at Objectives

Keep *Objectives* in mind when you draft *Results and Discussion*. Your writing is centered on the objectives. Therefore, you only present the results and discussion that help you achieve *the* objectives at the end of the *Introduction*.

If you happen to obtain valuable results that are unrelated to your objectives, you should save them for another publication or create another division if it is a long document like thesis or book. Mixing irrelevant ideas reduces the clarity of your writing and may confuse your readers.

7.3 RECA Approach to Results and Discussion

Key techniques for drafting *Results and Discussion* include reiteration, explanation, comparison, and analysis. They are all centered on the research objectives:

1. *Reiteration* integrates visuals into the text.
2. *Explanation* explores the reasons for the data as they are.

© The Author(s), under exclusive license to Springer Nature Switzerland AG 2022
Z. Tan, *Academic Writing for Engineering Publications*,
https://doi.org/10.1007/978-3-030-99364-1_7

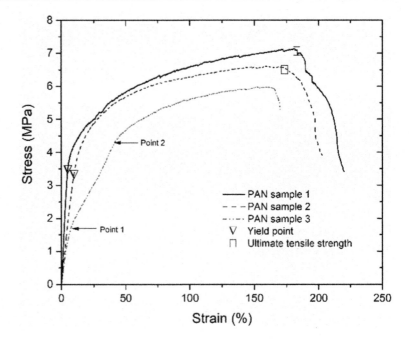

Fig. 7.1 Data validation (stress-strain curves of membranes; used with permission of Y Li ©2021)

3. *Comparison* compares your results internally or with those in the literature.
4. *In-depth analysis* deduces new knowledge aiming at the *objectives*.

The boundaries between these elements are often unclear. For example, comparing your own data in the same figure is part of integration, but comparing yours with those in the literature to support your argument is part of in-depth analysis. In addition, some journals require separating results from discussion, but the writing techniques remain the same.

Reiteration is only the first step in presenting results; it is not discussion yet. However, many junior researchers stop at this point – they either move on to another result or jump to drawing conclusions. Skipping *explanation, comparison,* or *in-depth analysis* is a typical mistake in poor writing. As a result, you cannot draw conclusions based on convincing argument. Furthermore, the results without in-depth analysis read like a technician's test report. Data alone do not contribute to new knowledge or help accomplish the objectives of the research.

Examples 7.1, 7.2, and 7.3 show the *Results and Discussion* section written by native English speakers (Fuller et al. 2012). The column on the right analyzes the writing corresponding to *reiteration, explanation, comparison, in-depth analysis,* and so on.

7.3.1   Reiteration of Results

Example 7.1 shows effective integration of visual into text. The text reiterates the legends, labels, results, and so on in the figures, which are clearly shown in the figures. The readers can capture the main ideas by looking at either the visuals or the text only.

Example 7.1. Reiteration by Integrating Visual into Text

Original writing	RECA analysis
......Figure 4 shows <u>the O/C and N/C ratios of the most intense peaks in the mass spectra as a function of their molecular mass from both stages. For clarity</u> only the <u>most intensive 200 peaks in the mass spectra are shown here,</u> which correspond to about 80% of the total ion intensity. The molecular formulas from all 20 extraction points were combined and their intensities from each point summed. ...... <u>Stage 3 is shown in blue and stage 8 in red,</u> superimposed on top of the stage 3 data. <u>The elemental ratios of O/C and N/C ratio were found to vary with particle size fraction (between the stages) and with molecular mass range.</u>	*Reiterate the y-axis and x-axis.* *Key information not shown in the figure* *Reiterate the legends.* *You can see the same information from the figure alone.*

	Figure shows most information described in the text.

| Figure 4 Elemental ratios of formulas for the most intense 200 peaks in the mass spectra in both stages (3 and 8) as a function of their neutral mass: (a) O/C ratio; (b) N/C ratio. Symbol size reflects peak intensity. (Reprinted with permission from ACS ©2012) | *Figure caption title further enhances clarity of writing.* |

To emphasize, you need to clearly explain the content in any relatively complex visual. A basic requirement in academic writing is that your readers can understand most of the results *either* by looking at the figure *or* by reading the text only. They should be relatively independent from each other.

7.3.2   Explanation and Comparison of Results

Example 7.2 shows presentation of results by explanation and comparison before in-depth analysis. Typical keywords introducing these elements include *the reason is, due to, in comparison,* and the like. All these elements are part of discussion.

7.3.3   In-Depth Analysis

In-depth analysis is essential to writing the *Results and Discussion*. It ensures the accomplishment of objectives. In-depth analysis is especially important to peer-reviewed publications, such as journal articles, which are aimed at advancing science and technology. Therefore, many reviewers reject manuscripts submitted to international journals on the ground of lacking in-depth analysis.

As seen in Example 7.3, you can conduct in-depth analysis by comparing your results with those in the literature, exploring the mechanisms behind an observation, deducing new knowledge from the known, presenting limitations of your work, and so on. Example 7.3 begins with the method used for in-depth analyses, followed by deducing new knowledge related to organic compounds, and ends with limitations of the work (*due to complexity*).

Example 7.2. Explanation and Validation of Results

Original writing	RECA
......Figure 4a <u>shows</u> 15% more compounds containing only C, H, and O among the 200 most abundant species in stage 3 than stage 8 (165 compounds compared to 141). For these CHO species in stage 3, 69% of the top 200 peaks by number and 87% by intensity have a neutral mass of <200 Da. <u>In comparison</u>, for stage 8 only 57% by number and 45% by intensity are in this mass range. <u>This is due to</u> a larger number of compounds in stage 8 with no oxygen content, but only C, H, and N, <u>indicating the presence and importance of</u> amine type species in this submicron particle size range. The similarity of the N content between the stages <u>is due to</u> this increased abundance of CHN compounds in stage 8 being balanced by presence of more oxidized nitrogen species (CHNO) in stage 3. Figure 4b <u>shows that</u> the nitrogen content is largely similar for both stages; <u>however</u>, for stage 3 for a neutral mass of <200 Da a large number of high-intensity compounds are present on the baseline with an N/C ratio of zero, <u>indicating that</u> they contain only C, H, and O, which is accordance with the higher O/C ratio in the bulk composition in the same mass range (Table 1).	*Reiteration and integration of visual into text* *Comparison of internal data* *Explanation* *Simple analysis* *Explanation* *Reiteration by integration* *Explanation* *Simple analysis*

To emphasize, you should always attempt in-depth analysis in academic writing. The discussion begins with the most important and the most relevant to the objectives, followed by less important ones. This part of writing begins with a clear statement followed by logical arguments; avoid assumption that your readers will understand or accept your ideas without evidence.

You can compare your own results with those in the literature and explain whether they are the same or different. If your findings are different from others, they deserve even further analysis. First, double-check your methodology to exclude any possible mistakes in data collection and analysis. (*See* "Abstract") Then, you can claim that your findings are new. Finally, support your claim with external and your internal results.

Example 7.3. In-depth Analysis of Results

Original writing	RECA
For all compounds below m/z 150 (181 compounds) in the large size fraction (stage 3) the N/C ratio is clearly correlated with the overall particle loading and wind speed (Supporting Information Figure S2). This strongly suggests that compounds in the low-mass region with high nitrogen content may be soil derived. Soil organic compounds are known to have a large number of nitrogen-containing compounds. The sampling location close to a dairy farm could also partly explain the high N/C ratio of the large particle fraction under high-wind conditions, due to the high N content of animal waste. This correlation is not as strong if all compounds up to m/z 350 are considered.	*Analysis by correlation* *Use SI for clarity.* *Deducing new knowledge* *Support argument with results in the literature.*
As shown above it is often difficult to identify groups of compounds originating from a common source, and correlations of peak intensities over time allow the possible identification of compounds with a common or similar atmospheric history. Thus, for each identified formula the variation of intensity across the 20 sampling points was compared to every other formula in the entire data set. Linear regression analysis was used to indicate correlations, and any correlation with an R2 value of greater than 0.8 was defined as statistically significant. This analysis assumes that matrix ionization effects are constant throughout the RDI stripe due to the hydrocarbon grease applied to the Mylar strips. This resulted in a number of sets of correlating species for stage 3. Three of these correlating sets of species are shown in Supporting Information Table S1, which gives for each set the formula and the total peak intensities across all 20 extraction points.	*Further analysis of the correlation* *Further analysis of the correlation* *Comparison for analysis* *Analysis by linear regression* *Criterion for "significant"* *Assumption for analysis*
Figure 7 shows the variation in the averaged intensity for all the species in each set for each extraction point, which was subsequently normalized to the maximum average intensity extraction points. Series 1 contains both CHNO and CHN species, and series 2 contains only CHO compounds. Both series show a negative correlation to the brown-colored regions in stage 3 (Figure 2a). In contrast, series 3 contains CHNO compounds only and shows a positive correlation with the brown deposit (extraction points 9/10 and 17, Figure 2a). Series 3 also correlates with the overall N/C ratio and wind speed as shown in Supporting Information Figure S2, again suggesting that these compounds may be soil-derived and from local (farm) sources. No series of correlating species were identified by this method in stage 8 indicating a more complex composition and atmospheric history of the small-particle fraction.	*Figure 7 produced for in-depth analysis. Right on the objective "Comparison"* *New knowledge aiming at objectives* *Limitations of the work due to complexity*

7.3.4 Presenting Limitations

Presenting the weakness of your results, especially their uncertainties and errors, does not devalue your contribution to the field. Instead, you gain respect from your peers and readers who highly value professionalism and clarity in writing. It also demonstrates that you clearly understand the subject from both sides and present them objectively.

Example 7.4 is a direct quotation of one paragraph in the PhD thesis by Givehchi (2015). The first sentence clearly states that there are uncertainties and limitations in the research. Other researcher can decide whether to tackle these challenges.

Note

Explicitly stating an opinion is considered inappropriate or unnecessary in some cultures, but technical readers a clear point of view for engineering publication in English. However, the statement should be followed by a convincing argument.

Example 7.4. Presenting Limitations After Analysis

"In addition, <u>there are other uncertainties and limitations in this work</u>. First, in general, the following factors were excluded from the model: shape and morphology of nanoparticles, smoothness of the surface, oblique impaction of nanoparticles, and limited database for nanomaterial properties. <u>Second,</u> the calculation of capillary force can also be further improved. The relation between the capillary force and relative humidity for sub-30 nm particles at relative humidity of smaller than 10% was calculated using extrapolating of data by Pakarinen et al. (2005). We assumed zero capillary force at dry condition, which may cause some errors. <u>In addition,</u> the capillary force was calculated for spherical particles; however, the contact angles and the meniscus shape could affect the capillary force strength and should be considered in future work. Neither NaCl nor WOx particles used in our experiments is spherical, and the actual pull-off force based on capillary force may be smaller than the calculated values based on the assumption of spherical particles."

7.4   Quantitative Results

A basic requirement in writing *Results and Discussion* is providing only objective results with quantitative evidence; the evidence can be your own results or data in the literature. In contrast, qualitative subjective comments on the results without evidence undermine the value of your work.

Examples 7.5 and 7.6 compare quantitative objective descriptions with qualitative subjective comments on results. Qualitative writing should be either replaced or followed by quantification to improve clarity. Avoid using vague words like *good, bad, fast*, and *slow* to describe results. Instead, quantify them with measurements. (*See* Sect. 12.7.2.)

Example 7.5. Avoid Vague Subjective Description

Vague: Figure 5 shows that the model <u>agrees well with</u> the experiments.

Clear: Figure 5 shows that <u>the maximum difference</u> between the model and experimental data <u>is 0.5%.</u>

Clear: Samples are mounted onto a movable sample stage, and a <u>small amount of solvent (1–10 µL)</u> is then dispensed from a pipet tip on the surface of the sample.

Example 7.6. Avoid Statement Without Evidence

Rough draft: Filtration-based enrichment is characterized with <u>low cost and simple operation</u> without labeling process. A transparent membrane enables direct observation of the captured cells, and cells can be <u>easily</u> released by applying a reverse flow. However, some of the WBCs are also retained because of the occasional size overlap between CTCs (down to 10 µm) and WBCs (up to 20 µm). Moreover, the potential damage of CTCs and the clogging of filters should be considered in filter design and processing rate control.

Suggested revision: Filtration-based enrichment is characterized with low cost and simple operation without labeling process. [<u>Support the argument with evidence: for example, data in the literature to show the *low cost*, comparison between filtration-based enrichment</u>

and other technologies-based enrichment to show the level of complex in operation.] A transparent membrane enables direct observation of the captured cells, and cells can be ~~easily~~ released by applying a reverse flow. However, some of the WBCs are also retained because of the occasional size overlap between CTCs (down to 10 μm) and WBCs (up to 20 μm). Moreover, the potential damage of CTCs and the clogging of filters should be considered in filter design and processing rate control.

7.5   Precise Results and Discussion

Multiple sections in this book emphasize preciseness in engineering publications. Example 7.7 compares general with precise writing for *discussion* (Khodabakhshi et al. 2021). The words *behavior* and *properties* in this example are overly general in the context of writing. Describing the *behavior* and *change* with specifics improves the clarity of the writing.

Example 7.7. Writing with Precise Details

General: The <u>behavior</u> of the coefficient of restitution for both high- and low-energy substrates was explained by three mechanisms of translational kinetic energy dissipation of the colliding nanoparticles: particle-substrate interaction energy, particle-water interaction energy, and plastic deformation of the nanoparticle.

Precise: The <u>change of the coefficient of restitution with respect to water layer thickness</u> for both high- and low-energy substrates was explained by three mechanisms of translational kinetic energy dissipation of the colliding nanoparticles: particle- substrate interaction energy, particle-water interaction energy, and plastic deformation of the nanoparticle.

General: It was shown that the <u>collision properties</u> of 5-nm particles on a wet substrate depends on the thickness of the condensed water layer on the substrate, surface energy of the substrate, and impact velocity.

Precise: It was shown that the <u>coefficient of restitution</u> of 5-nm particles on a wet substrate depends on the thickness of the condensed water layer on the substrate, surface energy of the substrate, and impact velocity.

7.6   Engaging Writing

Another trait in presenting *Results and Discussion* is engaging your readers. Writing is for the communication between you and your readers. You draft as if they were sitting right in front of you. Thus, you should anticipate challenging questions from your readers when they read your document. Then write down the answers to those important questions, which can improve the clarity of your writing. Like any part of your writing, you should present *Results and Discussion* with active voice when you can. For example:

Example 7.8. Presenting Results with Active Voice

Passive: In Figure 2, smooth Bi_2WO_6 nanosheets <u>were observed</u> in the 25-BWO sample without other phases, and they aggregated to form irregular bulky nanosheet clusters.

Active: Figure 2 <u>shows</u> smooth Bi_2WO_6 nanosheets in the 25-BWO sample without other phases, and they aggregated to form irregular bulky nanosheet clusters.

7.7   Erratic Writing

Writing mistakes may appear in *Results and Discussion* section when you neglect one or more of *reiteration, explanation, comparison*, and *in-depth analysis*. The exact errors depend on the writer's training and experience in writing, and the following often appear in engineering publications:

- Poorly integrate visual into text.
- Present results without in-depth analysis.
- Interpret data in visuals by mistake.
- Forget error bars in results, if applicable.
- Lack convincing argument in *discussion*.
- Refuse to accept limitations.
- Contents that do not belong.

Working on the practice problems should help you understand these mistakes and avoid them in your writing.

7.8   Practice Problems

Question 1:   This writing shows poor integration of visuals into text. Identify the writing mistakes in the *Results and Discussion* and suggest corrections. (*Draft manuscript reviewed for the article by Zhao* et al. *(2013); published version may be different than what appears here.*)

[Beginning of the paragraph] Figure 6 shows the sketch of the room and the positions for eight measuring poles (*citation*). Heat sources, including human simulators, lights, computers, are present in the real office as well as in the simulations. Contaminant sources are the human simulators. Table 1 summarizes the parameters of the supply and return air including inlet temperature, outlet temperature, inlet velocity and outlet velocity. *[End of the paragraph]*.

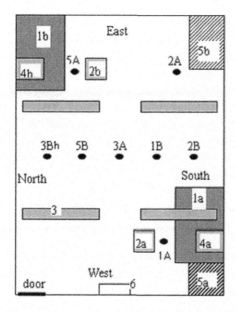

1 — table (a, b), 2 — human simulators (a, b), 3 — lights,

4 — computers (a, b), 5 — cabinets (a, b), 6 — displacement diffuser

Figure 6   (Reprinted with permission from Elsevier © 2009)

Table 1   Supply and return parameters

	Displacement	Grille
Inlet temperature (°C)	15.88	18.49
Outlet temperature (°C)	24.80	24.16
Inlet velocity (m/s)	0.3	1.55
Outlet velocity (m/s)	0.29	0.47
Ventilation rate (m³/s)	0.0562	0.0674

Question 2: This is an example of *results* without in-depth analysis. Identify the writing mistakes and suggest corrections.

Fig. 8 shows the percent changes in the aerodynamic force coefficients and rolling moment coefficients due to rain impact for various rain intensities with certain yaw angle. Different colored bars shown in the figure represent the percent changes in coefficients under different rain intensities. From Fig. 8(a) ~ (b), it is clearly that both the percent changes in coefficients of aerodynamic drag force and side force are in the character of ascent as the amount of rainfall is up for all the yaw angle calculated in this work. Fig. 10(a) indicates that the drag coefficients increases fall in the range from 1.5% to 38.6% across the yaw angle. And the relative high drag coefficients increases occur at large yaw angle: when the yaw angle equals to 30.97° or 35.76°, the drag coefficients increases are within the range of 13.0–39.0%. Fig. 8(b) indicates that the percent changes inside coefficients vary from 0.5% to 9.2% across the yaw angle. Similar to those of drag coefficients percent changes, the relative high drag coefficients increases occur at large yaw angle: when the yaw angle equals to 35.76°, the side force coefficients increases have a value between 2.5% and 10.0%. Fig. 8(c) shows that there is not a generalized regularity between the rain intensity and the lift force coefficients across the yaw angle range. However, distinct regularity still can be seen when the yaw is very low or relatively high: for the low yaw angle which equals to 6.85°, the percent changes in lift force coefficient has an upward trend with the increasing of rain intensity; on the contrary, for the high yaw angle which equals to 35.75°, the percent changes in lift force coefficient fall down as the rain intensity increasing. For the rolling moment coefficients, they increase as the rain intensity gets heavier across the yaw angle range as shown in Fig. 8(d). And the rolling moment coefficients increases from 1.8-10.0% are derived for the various rain fall rates. Furtherly, the relative high rolling moment coefficients increases occur at low yaw angle: when the yaw angle equals to 6.85°, the rolling coefficients reaches 4.0–10.0%.

Fig. 9 shows… *[Last paragraph in the same style.]*

Question 3: This is an example of incorrect interpretation of data in visual. Identify the writing mistakes and suggest corrections.

[Beginning of the paragraph] From Fig. 3, it can be seen that the benzene removal efficiency increased with increasing discharge power both in NTP or CPMC. The influence of discharge power on benzene removal efficiency can be ascribed to the influences of voltage and current……

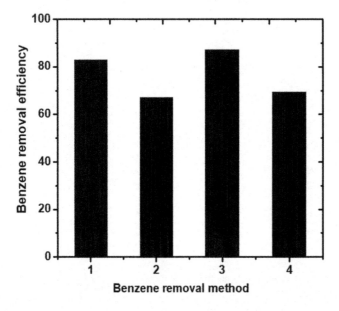

Fig. 3 The effect of discharge power on benzene by plasma only and plasma-MnO$_2$ systems. 1.NTP, 9W; 2. NTP, 4.5 W; 3. CPMC, 9 W; 4. CPMC, 4.5 W (reprinted with permission from Elsevier©2015)

Question 4: This is an example of contents that do not belong to *Results and Discussion*. Identify the writing mistakes and suggest corrections.

2.   Experimental

2.3   Characterization

Morphology was observed using a field emission scanning electron microscopy (FE-SEM, Tescan MAIA3 LMH) and transmission electron microscopy (TEM) in a bright field…

3.   Results and Discussion

3.2   Morphology and Microstructure

The morphology and microstructure of the samples were observed by FE-SEM and TEM techniques. Figure 2a shows that both flower- and rod-like shapes appeared in the 0-BWO sample, which may correspond to Bi_2WO_6 and $Bi_6O_6(OH)_3(NO_3)_3 \cdot 1.5H_2O$.

Question 5: Identify the sentences that do not belong to *Results and Discussion* and suggest corrections.

3.5   Optical Properties

[Beginning of Section 3.5] Excellent light absorption and appropriate band gap are important to strong photocatalytic activity. The optical properties of 25-BWO, 50-BWO, 75-BWO and 100-BWO were detected by UV-vis diffuse reflectance spectra.

Figure 5. UV-vis DRS of different samples: (a) the relationship between absorption and wavelength and (b) transformed by Kubelka–Munk method

Figure 5 shows that ….

Drafting Conclusions

8

8.1 Conclusion Meeting Objective

The main body of a typical academic publication begins with *Introduction* and ends with *Conclusions*. *Introduction* serves as a framework into which readers understand the background, the motivation, and the overview of the work. *Conclusions* tie the main objectives together; you convince your readers that you have accomplished the *objectives* as stated at the end of *Introduction*.

The *Conclusions* section of a document answers the ultimate question: have you accomplished the *objectives*? The conclusions, which are drawn from the results and in-depth analyses, point all the evidence at the key findings and new knowledge. You can also include recommendations to your readers, such as a course of action for further research.

Nonetheless, the conclusions must be drawn from the evidence that is presented in the same document. Avoid in-text citations in the *Conclusions* section. A well-written *Conclusions* section is closely linked to the *Objectives* presented earlier. Your task at this point is to reiterate the main points in the objectives with answers.

Example 8.1. Well-Written *Conclusions*

[First paragraph of the Conclusions] We have <u>developed a new technique</u> to analyze the organic composition of atmospheric aerosol on a molecular level with a high time resolution by combining a new extraction and ionization technique, LESA, with RDI samples and <u>ultrahigh resolution mass spectrometry</u>. LESA facilitates analysis without time-consuming sample preparation or extraction and also allows direct extraction from a sample surface with <u>high spatial resolution</u>, which when combined with RDI samples results in highly time-resolved information of <u>ambient aerosol organic content</u>.

[Second paragraph of the Conclusions] Ultrahigh-resolution mass spectrometry was used to determine the <u>chemical composition of organic compounds</u> and to investigate their changes in intensity over the sampling period in two aerosol size ranges. <u>Changes in the</u> chemical composition could be related to changes in meteorological <u>conditions</u>; for example, the signal intensity of a subset set of nitrogen- containing compounds was found to be associated with the local wind speed suggesting that <u>these compounds are largely of local origin</u>. Groups of compounds were identified that showed similar trends in abundance over the sampling period indicating that these

---

© The Author(s), under exclusive license to Springer Nature Switzerland AG 2022
Z. Tan, *Academic Writing for Engineering Publications*,
https://doi.org/10.1007/978-3-030-99364-1_8

compounds originate from the same source. This was demonstrated by the identification of homologous series, which are likely associated with biomass burning sources. A further characterization of the structures of the compounds, e.g., by performing MS/MS experiments, was not possible due to the generally low peak intensities. This could be overcome in future studies by a lower rotation speed of the RDI during sample collection, which would allow for higher sample loadings and might allow sufficient peak intensities to perform MS/MS experiments.

[Last paragraph of the Conclusions] While mass spectrometry is widely used to identify organic aerosol content, the combination with LESA and RDI sampling allows a much higher temporal sampling resolution than would be possible with other off-line techniques allowing one to identify marker compounds and to follow atmospheric processes in detail.

Example 8.1 shows the conclusions drawn from the results and in-depth analyses by Fuller et al. (2012). The first paragraph of the *Conclusions* is related to the first objective, developing a new technique (*see* Example 5.8). The second paragraph echoes the second objective, characterizing the aerosol particles, etc. (*see also* Example 5.8). These two paragraphs confirm that the researchers accomplished their objectives.

On the other hand, the last paragraph in the *Conclusions* section seems to be isolated from the rest of the paper. The authors might hope to address one reviewer's comments or concerns, or they might want to emphasize the practical value of their new technology. In terms of writing, however, this is a single-sentence paragraph without body evidence. This paragraph could be merged with another that has the same controlling idea.

8.2   Practice Problems

Question1: Following the guide in this book, identify the writing errors in and suggest revisions to the following *Conclusions* section.

1. In conclusion, we have demonstrated a novel mild two-step method to produce high quality rGO. After the reduction, the sheet resistance and work function of rGO decreased much. With different content functional groups and defects in the graphene lattice, sheet resistance and work function are highly influenced. RGO with the tunable work function can be applied in cathode or anode in optoelectronic devices.
2. In summary, it was shown that the collision properties of a nanoparticle to a substrate depends on relative humidity. Depending on the wettability of the substrate as well as the impact velocity, humidity can decrease or increase the adhesion of the nanoparticle to the substrate. In the collision of a nanoparticle with a high surface energy substrate, humidity reduces the adhesion of the particle to the substrate. However, for low surface energy substrates, the effect of humidity at high and low impact velocities is different. When the impact velocity is low, particle adhesion to the surface decreases with humidity. While for high impact speeds, the adhesion of the particle to the substrate with humidity can be reduced. Since the collision properties of nanoparticles to the substrate is determined by the particle impact energy dissipation, the mechanisms of energy dissipation in dry and wet impact were compared with each other. Recent studies show that using surfaces with high wettability deteriorates the filtration efficiency of the nanoparticles as well as bacteria. Moreover, bacterial adhesion to super-hydrophilic surfaces is very low. Due to the unavoidable presence of humidity in both of these phenomena, humidity can be considered as a factor to reduce the adhesion of nanoparticles and bacteria to surfaces with high wettability.

Drafting Other Sections

9

9.1 Drafting Title

Title, abstract, and keywords are three key elements that can improve the discoverability of your work. Without them, your work may only reach a small portion of potential readers in the world.

The title of a document is important to both the writer and readers. An effective title condenses the scope and purpose of the writing into one concise clause. It is expected to call for readers' attention and interest. The titles should be concise and precise. You should avoid the following writing errors in title:

- Redundant words
- Abbreviation
- Chemical formula
- Sentence form
- A rhetorical question

Example 9.1 highlights that words like *On*, *Studies on*, and *A Report on* are redundant for title. You can see many similar titles in published works.

Example 9.1. Redundancy in Title of Journal Article

Redundant:	A study of the effect of nitrogen dioxide on the absorption of sulfur dioxide in wet flue gas cleaning processes. *[Source: Siddiqi et al., 2001 Industrial & Eng Chemistry Res 40: 2116–2127.]*
Concise:	Effect of nitrogen dioxide on the absorption of sulfur dioxide in wet flue gas cleaning processes.

However, the entries considered redundant in the preceding list are acceptable for *subtitles* because they specify the purposes and scopes of the writing. More can be included in a subtitle because that is what subtitle is for. Title and subtitle are separated by a colon mark. *See* Example 9.2 (*source: search*

© The Author(s), under exclusive license to Springer Nature Switzerland AG 2022
Z. Tan, *Academic Writing for Engineering Publications*,
https://doi.org/10.1007/978-3-030-99364-1_9

in Google Scholar). More and more books have subtitles now, but most journal articles omit subtitle for conciseness.

Example 9.2. Titles and Subtitles

Book title:
- Nanoengineering: Global Approaches to Health and Safety Issues.
- Handbook of Polymer Nanocomposites for Industrial Applications.
- Cognitive Informatics, Computer Modelling, and Cognitive Science: Volume 1: Theory, Case Studies, and Applications.
- Air Pollution and Greenhouse Gases: from Basic Concepts to Engineering Applications for Air Emission Control.

Article title:
- Selection and interpretation of diagnostic tests and procedures.
- Coupling and entangling of quantum states in quantum dot molecules.
- Efficient bipedal robots based on passive-dynamic walkers.

As another example, the United States Environmental Protection Agency (US EPA) publishes annual reports on new light-duty vehicle greenhouse gas (GHG) emissions, fuel economy, technology data, and auto manufacturers' performance in meeting EPA's GHG emissions standards. The annual report published for 2019 has a title of *The 2019 EPA Automotive Trends Report* with a subtitle of *Greenhouse Gas Emissions, Fuel Economy, and Technology since 1975* (EPA 2020). The subtitle contains much more information than the title itself.

9.2   Drafting Authorship

9.2.1   Corresponding Author

The corresponding author is typically the one who oversees the work and who has a permanent position where the work was produced. The corresponding authors of works produced at universities are normally the academic supervisors of the research teams instead of the students. The students often leave the universities after completing the research works. Additionally, students may not have the depth of knowledge to respond to readers' inquiries.

9.2.2   Authorship Order

Consult with your superior before finalizing the order of authorship. It may be different from what you have learned before. In many disciplines and countries, the authorship order indicates the weight of contribution. Most agree that the first author contributes most to the work and that the last author representing predominantly the one who oversees the entire work. However, the most senior may be listed as the second author and the corresponding author in many countries.

Example 9.3 shows the title and authorship of the sample article written by Fuller et al. (2012). The sample paper was written by native English speakers. More examples from the same reference are used as well-written examples throughout this book.

Example 9.3 Title and Authors of the Sample Paper

Direct surface analysis of time-resolved aerosol impactor samples with ultrahigh-resolution mass spectrometry.

Stephen J. Fuller[1] Yongjing Zhao[2] Steven S. Cliff[2] Anthony S. Wexler[2] Markus Kalberer*[1].
[1]Department of Chemistry, University of Cambridge, Lensfield Road, Cambridge CB2 1EW, U.K.
[2]Air Quality Research Center, University of California–Davis, Davis, California 95,616, United States.
*E-mail: markus.kalberer@atm.ch.cam.ac.uk

9.2.3 Affiliations

Most publications list affiliations right after the names of co-authors. Make sure that the affiliation is the institution where the co-author made the contributions to the work instead of the paper – the paper and the work are different concepts. For example, if a university student is employed by a company after successful submission of the dissertation, the university instead of the current employer should be listed as the affiliation when a journal article is written based on the student's thesis work.

Example 9.3 shows the title, authorship, and affiliations of an article co-authored by my former graduate student, a collaborator in another country, and myself (Givehchi et al. 2018). The first author (Givehchi) carried out the research at the University of Waterloo under the joint supervision of the second author (Li) and the last author (Tan). They all have major contributions to the research work or the writing of the article or both. At the time of publication, Givehchi was employed by the University of Toronto. However, the University of Toronto cannot be listed as Givehchi's affiliation in the article. Z. Tan was affiliated with both institutions.

Example 9.4. Title, Authorship, and Affiliations

Filtration of sub-3.3 nm tungsten oxide particles using nanofibrous filters

Raheleh Givehchi[1], Qinghai Li[2,*] and Zhongchao Tan [1,2,*]
[1]Department of Mechanical & Mechatronics Engineering, University of Waterloo, Waterloo, ON N2L 3G1, Canada; raheleh.givehchi@utoronto.ca
[2]Tsinghua University—University of Waterloo Joint Research Centre for Micro/Nano Energy & Environmental Technologies, Tsinghua University, Beijing 100084, China.
*Correspondence: liqh@tsinghua.edu.cn (Q.L.); tanz@uwaterloo.ca (Z.T.)

9.3   Drafting Abstract

The abstract of your publication allows your readers to decide whether it is necessary to continue reading beyond this point. The abstract should be treated as a stand-alone document, and it should tell a complete story apart from the original document.

The length of an abstract varies with the document and publisher. It is typically 200–250 words for an article and about 1000 for a thesis. Extended abstracts, normally three to five pages each, are collected in proceedings of international conferences. Regardless of their lengths, all abstracts follow similar principles.

There are two types of abstracts, *descriptive abstract* and *informative abstract*. The former briefly summarizes the objectives and methodology of the research. The latter expands the descriptive abstract with more information like results, conclusions, and recommendations. You need to use descriptive abstracts in compiled documents, such as a collection of conference papers and technical reports of a large project before the final one is ready. Informative abstracts work best for a wide range of readers who are likely interested in the conclusions and recommendations.

A well-written abstract follows *Why-How-What* pattern. It contains three key elements: objective (*Why*), methodology (*How*), and results or conclusions (*What*). A short abstract with word limit should begin with the subject and the scope of the work. However, many writers, including native English writers, often precede the topic sentence (objectives) with background information. As a result, the abstract lacks sufficient information on results or conclusions (i.e., *What*).

Example 9.5 shows a well-written abstract (despite its imperfection) that follows the *Why-How-What* pattern (Fuller et al. 2012). The abstract begins with the *Why* [the research is important and necessary]: uncertainties exist, composition matters, aerosol particles are poorly characterized, etc. Then it introduces *How* [to address the knowledge gaps]: a new technique. Finally, the abstract ends with *What* [are the findings]: groups of compounds with common sources.

Despite its logical structure, this abstract deserves further refinement. It could have begun with a shorter *Why* and ended with a longer *What*. The *Why* part has 67 words; the *What* part has about 34 words and it lacks details. The word limit for *Abstract* in *Analytical Chemistry* is 80–200 words. Thus, the authors could have enhanced the clarity of writing by adding 45 more words into the *What* portion.

Example 9.5. A Well-Structured Abstract

Original writing	3 W's analysis
Aerosol particles in the atmosphere strongly influence the Earth's climate and human health, but the quantification of their effects is highly uncertain. The complex and variable composition of atmospheric particles is a main reason for this uncertainty. About half of the particle mass is organic material, which is very poorly characterized on a molecular level, and therefore it is challenging to identify sources and atmospheric transformation processes. We present here a new combination of techniques for highly time-resolved aerosol sampling using a rotating drum impactor (RDI) and organic chemical analysis using direct liquid extraction surface analysis (LESA) combined with ultrahigh-resolution mass spectrometry. This minimizes sample preparation time and potential artifacts during sample workup compared to conventional off-line filter or impactor sampling. Due to the high time resolution of about 2.5 h intensity correlations of compounds detected in the high-resolution mass spectra were used to identify groups of compounds with likely common sources or atmospheric history.	*Why: uncertainty, composition, organic matters, poorly characterized, technical challenge* *How: a new technique with high time resolution* *What: compound detected in spectra; source*

Writing an excellent abstract is challenging, even to the native English speakers. Therefore, it is normal for the non-native English speakers to make some mistakes in writing an abstract. Examples 9.6 and 9.7 show two abstracts written by non-native speakers, and they illustrate some of the typical mistakes that non-native English writers might make.

Example 9.6. Poorly Written Abstract (1)

The particle number size distribution (3 nm–10 μm) was conducted in May of 2011 at the urban sampling site of 18 City. Their pollution characterization and their dependencies on gaseous contaminants and meteorological factors were synchronously investigated. The diurnal average total of particle number were 568 cm^{-3} in nucleation mode (3–20 nm), 14,909 cm^{-3} in Aitken mode (20–100 nm), 7418 cm^{-3} in accumulation mode (0.1–1 μm) and 2 cm^{-3} in coarse mode (1–10 μm), respectively. The particle number, surface area and volume size distributions respectively presented the single peak, double peak and four peaks pattern. The diurnal variation of particle number in nucleation mode was mainly influenced by nucleation events in spring. The diurnal variation of particle number in Aitken mode was closely correlated with the traffic densities. The maximum contribution ratios to the total of the particle number, surface area and volume concentration were particles in Aitken mode, accumulation mode and accumulation mode, respectively. ......The absolute values of correlation coefficient between other factors (SO_2, NO_2, PM_{10}, PM_1 and visibility) and the particle number concentration was maximum in accumulation mode and indicating that the most notable method was decreasing the particle number concentration in accumulation mode for improving atmospheric environment quality. Local wind speed played an important role in shaping the particle number size distribution in the urban area of Jinan, China. With the increasing of wind speed, the particle number concentration increased in nucleation mode and decreased in the Aitken mode and accumulation mode. The particle number concentration in haze weather was evidently higher than that in clear weather. The union action of gaseous contaminants and meteorological factors was supposed to supply a favorable environment for the formation and growth of secondary particles to lead the lower visibility in haze weather.

Example 9.6 shows several errors in writing an abstract. It begins with one sentence for *Why* (the scope), followed by another sentence for *How* (the method). Both *Why* and *How* lack clarity. The *What* part of the abstract is long and presented at a rapid pace. This 293-word-long abstract exceeds the word limits of most journals. Finally, this abstract lacks conciseness and clarity. It becomes useless and boring to many readers.

Abstracts are expected to be written with clarity and conciseness. Avoid omitting articles or important transitional words and phrases. You can use acronyms after definition, but avoid in-text citations, equations, references, and the like. Example 9.7 shows an abstract with these errors (Du et al. 2006). Despite its correct structure, this abstract has many writing errors. The most obvious blunder is the presence of references, one enclosed by brackets and another by parentheses. There are 220 words in this abstract, and 68 words are for unneeded references. Furthermore, the authors misused intensifiers such as *significant, best, critical,* and *subtle,* which are subjective and vague words.

Example 9.7. Poorly Written Abstract (2)

In the computational fluid dynamics (CFD) modeling of gas–solids two-phase flows, <u>drag force</u> is the only accelerating force acting on particles and thus plays an important role in coupling two phases. <u>To understand</u> the influence of drag models on the CFD modeling of spouted beds, several widely used drag models available in literature were <u>reviewed</u> and the resulting hydrodynamics by incorporating some of them into the CFD simulations of spouted beds were <u>compared</u>. The results obtained by the different drag models were <u>verified</u> using experimental data <u>of He et al. [He, Y.L., Lim, C.J., Grace, J.R., Zhu, J.X., Qin, S.Z., 1994a. Measurements of voidage profiles in spouted beds. Canadian Journal of Chemical Engineering 72 (4), 229–234; He, Y.L., Qin, S.Z., Lim, C.J., Grace, J.R., 1994b. Particle velocity profiles and solid flow patterns in spouted beds. Canadian Journal of Chemical Engineering 72 (8), 561–568.]</u> The quantitative analyses showed that the different drag models led to <u>significant</u> differences in dense phase simulations. Among the different drag models discussed, the Gidaspow <u>(1994. Multiphase Flow and Fluidization, Academic Press, San Diego.)</u> model gave the <u>best agreement with</u> experimental observation both qualitatively and quantitatively. The present investigation showed that drag models had <u>critical and subtle impacts</u> on the CFD predictions of dense gas– solids two-phase systems such as encountered in spouted beds.

9.4 Drafting Executive Summary

Executive Summary is normally considered as part of the body of a long document. An executive summary of a long document has similar functions to an abstract of a short document. The main body of a long document begins with an executive summary. The executive summary may be published separately representing the original document. Thus, avoid cross-references in Executive Summary.

A well-written executive summary is a condensed version of the long document; it also requires clarity and conciseness in writing. It may contain objectives, methodology, results, conclusions, and recommendations, if applicable. Unlike an abstract, an executive summary may include essential figures, tables, or footnotes to maintain its independence from the original document.

An executive summary, with key divisions following similar order of sequence, is usually about 10% of the length of the original document. For example, the Executive Summary of *The (US) EPA Automotive Trends Report – Greenhouse Gas Emissions, Fuel Economy, and Technology since 1975* is 13 pages long, and the full report has 155 pages (EPA 2020).

9.5 Drafting Keywords

Keywords are usually listed right after Abstract or Executive Summary. Keywords are used for indexing by database and search engines to facilitate easy discovery of the document. With advances in online access to knowledge, most documents are now published online. Most search engines use keywords to determine the relevant documents for potential readers.

Keywords should be carefully selected. To choose the right keywordsyou can start with a list of terms and phrases that are used repeatedly in your document. Make sure that your list of keywords includes all key phrasesterms for procedurescommon abbreviationsand so on used in your research.

Then type your keywords into a search engine and check whether the results that show up match your keywords. This may be an imperfect solution but helps you determine whether the keywords in your publication are appropriate

9.6   Table of Contents

Table of Contents (TOCs) is useful to a long document, excluding journal articles and conference papers. A TOC enables quick review of the document structure because the TOC lists all major headings and their page numbers. A typical TOC contains one to three (or one to two) levels of headings. A TOC with more than four levels of headings would create visual clutter.

Most word processing software can create TOCs with numerous styles. You can also format the TOCs as you prefer.

9.7   List of Figures and List of Tables

List of Figures identifies the caption titles and locations (page numbers) of the figures (charts, drawings, photographs) in a long document. Articles in periodicals and proceedings of conferences normally do not use lists of figures.

Similarly, *List of Tables* identifies the caption titles and locations (page numbers) of the tables in a long document. The List of Tables allows readers to locate the tables of interest quickly and easily. Articles in periodicals and proceedings of conferences normally do not use lists of tables. *Table of Equations* is unneeded anywhere in the document.

> **Note**
>
> Only long documents need *Table of Contents*, *List of Figures*, and *List of Tables*. They can be created automatically using word processing software. You need to make sure that their font sizes and typefaces are compatible with those in the main document.

9.8   Abbreviations, Acronyms, and Symbols

Abbreviations and acronyms are frequently used in academic writing for engineering publications. An abbreviation is a shortened word, for example, *Dr.* for *doctor*. An acronym is a word created using the first letters of the words in a phrase, such as *DOE* for Department of Energy and *UW* for University of Waterloo. Proper abbreviations and acronyms enhance the readability and conciseness of the document they belong to.

Most abbreviations can be used without definition in the document because they are accepted by fellow professionals in the field and occasionally the public. For example:

- Social titles (*Mrs.*, *Mr.*)
- Professional titles (*Dr.*)
- Directions in city address (*SW* for south west)
- International units (*mm* for millimeter; *s* for second; *kg* for kilogram; *K* for Kelvin; *Pa* for Pascal)

Units of measurement are normally used in their abbreviated forms. The International System of Units (abbreviated SI) should be used for publications aimed at international readers (*see* Sect. 16.2.15. Formatting units and symbols). Dual systems of units are occasionally used in English documents, especially those traditionally published in the United States of America (USA).

Avoiding acronyms without definitions is important to the clarity of writing. You should define an acronym right after its first appearance in the text (*see* Example 9.8). You may redefine the same acronym for different chapters of a long document such as a thesis and book. It helps the readers who read only part of the long document.

Example 9.8. Defining and Using Acronyms

Research in <u>Geographic Information Systems (GIS)</u> is advancing rapidly… <u>GIS</u> is an increasingly important technology to the military.

For occasionally used abbreviations or acronyms in a long document, you need to reverse the definition (with the full text in parentheses) if it appears at a reasonable interval. (*See* Example 9.9.) Redefinition of acronyms enhances the clarity of writing by reminding the readers of their meanings because they may have forgotten how they are defined when they first appear.

Example 9.9. Reverse Definition of Acronym

- He is an expert in <u>GIS (Geographic Information Systems)</u>.
- EPA (Environmental Protection Agency) of the USA (United States of America).

Acronyms usually act as nouns in sentences. When an acronym follows an article (*a; an*) in a sentence, you need to choose the right article based on the sound of the acronym rather than the original phrase of the acronym, for example, *an unmanned aerial vehicle* (*UAV*) but *a UAV*. You can pluralize an acronym with addition of a lowercase *-s* (e.g., *UAVs*). Acronyms can also be made possessive with an apostrophe followed by a lowercase *-s* (e.g., *UAV's*).

Many professional organizations define their own acronyms. Therefore, you need to learn and use their acronyms correctly. For example, *ASME* is for *American Society of Mechanical Engineering*, but the *American Institute of Chemical Engineers* uses the acronym *AIChE* instead of *AICE*. As another example, *PE* and *PEng* are used for the same text of *Professional Engineer* in the USA and Canada, respectively.

Acronyms are subjects in sentences. They should be treated as singulars, which require singular verbs even though they stand for plurals. For example, AIChE *was* established in 1908.

Note

- Don't create abbreviations; instead, use widely accepted ones.
- Don't use periods in uppercase acronyms (AM, PM, NDA, IRA); use period marks for lowercase acronyms (a.m., p.m., e.g., etc., i.e.)
- Don't follow an abbreviation or an acronym with double period marks (see Punctuation).

9.9 Drafting Foreword and Preface

Foreword is an optional introduction of a long document, such as a book or a formal report. Journal articles, conference papers, or graduate theses do not have Foreword section.

The Foreword is normally written by an influential and important person with authority in the field to introduce the author(s) and the document to the readers. A nicely crafted foreword can also serve as an endorsement for the value of the document. Like a letter to the readers, *Foreword* is signed and dated at the end.

Preface is another optional introduction of a long document. Most books have Prefaces. *Preface* follows *Foreword* when both exist. It is written by the author(s) of the original document.

A preface introduces the contexts, including driving force behind the writing, motivation of the work, and purpose of the document. A typical preface also contains acknowledgments to those who have contributed to the preparation and publication of the work. However, Acknowledgment can be a stand-alone section located at the near end of a short document like as a journal article.

9.10 Drafting Acknowledgments

You can acknowledge your financial sponsor(s) and people who contributed to your publication but not significant enough to become the co-author(s). Example 9.10 is the *Acknowledgments* section in the article by Li and Tan (2020).

Example 9.10. Acknowledgments

The authors would like to acknowledge the financial support from the Natural Sciences and Engineering Research Council of Canada (RGPIN-2020-04687) and GCI Ventures Capital, Inc.

9.11 Drafting References and Bibliography

A list of references appears near the end of the document. Listing references in a consistent format allows readers to locate and access the works cited for further information on the subject. (*See also* Sect. 9.12.) It also allows the readers to validate the data used by the authors to support the conclusions in the original document. Equally important, it helps avoid plagiarism by giving proper credit to the original creators of the knowledge.

A bibliography is a list of all the sources that are used to support the writing. It usually includes references and additional resources for further reading by some readers who may need additional information. The sources include books, articles, theses, reports, creditable online sources, and other works. They are normally listed alphabetically by the authors' last names.

9.12 Appendices and Supplemental Materials

Supplemental materials, which are nonessential to the primary readers, are located at the end of the documents. They are labeled as *Appendices* in long documents and *Supplemental Information* (or the like) in short documents. Many prestigious journals require authors to use supplemental information because of page limit. The supplemental materials may include detailed derivation of equations, nonessential visuals like complex maps, coding for data analysis, and credentials of main researchers. Normally supplemental materials are only available online.

9.13   Glossary

A glossary is a list of definitions of terms. It is optional to a short document but may appear after *Bibliography* or *Appendices* to a long formal document. You need to alphabetically arrange the entries of a glossary for clarity. A glossary seems to have the same function as an index, but the glossary has much more information than the index.

9.14   Index

Index, if included, is located at the very end of a long document. It is a list of the keywords and phrases used in the document. Like bullet points, the index can also include subentries. Unlike glossaries, however, indices do not include definitions. Indices are available in most books but unneeded in articles, short reports, or theses.

You can start compiling the index when your final manuscript is ready. At this point, you read through the document from cover page to the end. Meanwhile, you note the key terms and their page numbers each time they appear. The entries in the index can include keywords and phrases in the table and figure caption titles. Finally, you need to alphabetically arrange the index entries with page numbers so your readers can locate them easily.

Manually compiling an index is a tedious task. Nowadays, software is available for this kind of tasks. For example, the American Society of Indexers (www.asindexing.org) lists software available to indexers for this process. Most word processing software, such as *Microsoft Word*, has this built-in function.

9.15   Practice Problems

Question 1: Highlight the unclear pronouns, references, and cross-references and suggest revisions to improve clarity of writing.

1. [Unclear pronoun] The method using functionalized nanostructure to capture CTCs have a strengthened affinity to CTCs. This is because the functionalized nanostructure can accommodate more ligands by its large surface area and has intimate interactions with CTCs. Based on this, microchannel embedded with functionalized nanostructure offers higher capture efficiency (typically ~60–95%) and purity with processing rate of around 1–2 mL/h (Table. 1).

2. [Ambiguous pronoun] The optimal ratio of antibody to aptamer was 1:300. In capture of 10–1000 CCRF-CEM cells from 1 mL whole blood, the prepared microchannel yielded capture efficiencies higher than 90% at a high flow rate of 7.2 mL/h. Beyond that, Nellore et al. [102] used multiple aptamers (S6, A9 and YJ-1) on the porous graphene oxide membranes to capture different kinds of CTCs.

3. [Ambiguous pronoun] If the magnetic beads are not detached from the CTCs after capturing, its high opacity may affect the observation of CTCs under optical microscope.

4. [Unclear cross-reference] The results stated above reveal that the choice of drag models can significantly affect the simulated flow patterns. It seems that both under- and over-estimations of drag coefficients in the dense phase bring about the deviation of the flow pattern from stable spouting operation, resulting in a bubble/slug flow pattern. For dense phase systems, van Wachem et al. (2001) showed that the Syamlal and O'Brien (1988) model led to lower predictions of pressure drop and bed expansion, in good agreement with our results. (Du et al. 2006)

5. [Unclear source] The heat dissipation of LIB is important to the battery cooling and thermal management. The excessive heat accumulation and uneven temperature distribution of LIB eventually cause the performance degradation. The heat conductivities of separator and electrolyte are much smaller than the other components of LIB. As a result, the thermal resistance within a 20 μm separator phase counts for about 70% of the total resistance in the battery radial direction. Thus, it is important to improve the thermal properties of battery separator. (Li and Tan. 2020)

Question 2: Revise these sentences with concise cross-reference.

6. Zhang et al. (2021) proposed a ligand-based microchannel capable of CTCs enumeration and purification.

7. *[Beginning of Introduction]* Ambient air pollutants, including particulate matter of aerodynamic diameter less than 2.5 μm and 10 μm (PM2.5 and PM10, respectively), ozone (O3), nitrogen dioxide (NO2), sulfur dioxide (SO2), and carbon monoxide (CO), are major environmental factors causing disease and death worldwide (Fenech et al. 2019; Fischer et al. 2020; Guo et al. 2019; Hanigan et al. 2019; Karimi et al. 2019; Lee et al. 2019; Pope et al. 2019; Wang et al. 2019; Wu et al. 2019; Zhang et al. 2019; Zhu et al. 2019). *[End of the paragraph] [Source: Hu et al. Environment International (2020) 144: 106018]*

8. Various studies have investigated the performance of nanofibrous filters fabricated with various polymers mounted in a cone-shaped filter holder in a laboratory-scale research (Bian et al. 2020, Cho et al. 2013, Givehchi et al. 2016, Liu, C. et al. 2015, Liu et al. 2020, Podgórski et al. 2006, Wang, N. et al. 2014, Yun, Ki Myoung et al. 2010, Yun, K. M. et al. 2007). The size-resolved filtration efficiencies reported in these studies varied from 0 to 100% for submicron particles; the efficiency greatly depended on the filter properties, including the number of nanofiber layers, fiber material, fiber diameter, filter porosity, filter basis weight, and filter thickness.

Question 3: Revise these titles following the guide in this book.

1. Study of a Novel Tornado-Like-Vortex Generator with Intelligent Controller
2. An electrostatic lens for focusing charged particles in a mass spectrometer
3. Experimental research on filtration performance optimization of airliner cabin air filters
4. On the kinetics of the absorption of nitric oxide into ammoniacal cobalt (II) solutions
5. ECG P wave abnormalities
6. A General Model for the Deposition of Entities (or Species) with a Wide Size Range from Molecular to Nano-sizes on the Nanofibers
7. A study of the effect of nitrogen dioxide on the absorption of sulfur dioxide in wet flue gas cleaning processes

Question 4:   Read the abstract of the sample paper: https://doi.org/10.1016/j.buildenv.2020.107392

1. Identity *Why-How-What* in the *Abstract* section.
2. Revise the abstracts to improve their clarity and conciseness.

- **Sample paper:** Zhang X, Liu J, Liu X, Liu C, 2021, Performance optimization of airliner cabin air filters, *Building and Environment* 187: 107392

Part III

Engineering Language Skills

10.1 Writing with Simple Language

A well-trained engineer uses plain language to explain a complex idea. This skill requires systematic training and continual practice. You should always use the simplest word for enhanced clarity for most readers and use specialized alternative when needed. Again, conciseness is important, but clarity is essential to engineering academic writing. For example, you can improve conciseness by omitting articles, pronouns, or verbs in sentences. However, it should be done without grammatical errors or loss of clarity.

Sometimes, it is challenging to decide the level of specialist vocabulary, especially when you have many choices. The best choice is to allow as many readers as possible to understand; you need to read widely in your field to understand the language that your fellow researchers use. Meanwhile, you should avoid big words and legal words in academic writing.

1. Avoid Big Words

Write in plain, ordinary English and your readers will enjoy reading it. Although formal language is used in academic writing, you may want to avoid important-sounding big words. Instead, use short, simple ones. Fancy language often frustrates the readers, especially non-native English speakers. For example, Table 10.1 compares some big word with simple ones.

Table 10.1 Big words vs. simple words

Big word	Simple word	Big word	Simple word
Ascertain	Find out	Incombustible	Fireproof
Assist	Help	Initiate	Start
Commence	Begin/start	Necessitate	Need
Demonstrate	Show	Render	Make
Endeavor	Try	Substantiate	Prove
Enquire	Ask	Terminate	End/stop
Erroneous	Wrong	Transmit	Send
Facilitate	Help	Utilize	Use

Z. Tan, *Academic Writing for Engineering Publications*,
https://doi.org/10.1007/978-3-030-99364-1_10

2. Avoid Legal Words

Avoid these legal words in technical writing for simplicity: *forthwith, hereof, thereof, henceforth, heretothereat, whereat, hereat, herewith, therein, whereon,* etc. Example 10.1 shows how to replace legal words with simple ones in engineering writing.

Example 10.1. Legal Words Vs. Simple Words

Vague: The <u>said</u> experimental condition ...
Clear: The experimental condition ...
Vague: The <u>aforementioned result</u> agrees with those in the literature.
Clear: A and B agrees with each other.

10.2   Writing with Right Pace

Pace is the speed of presentation. A carefully adjusted pace improves clarity and reduces ambiguity of your writing. Control your pace when you present your ideas to the readers. Pace can be controlled by simple sentences, smooth transition, and so on. Example 10.2 compares hastiness due to complex sentences with right pace using simple sentences. Hasty writing may also result from assuming that the readers are experts in the areas of research.

Example 10.2. Pace of Presentation

Hasty:
The electrospinning device is powered by a high-voltage power supply and produces nanofibers in the range of 10–800 nm in diameter. It is designed to operate under normal conditions of room temperature and low relative humidity, within a ventilation hood, and may be used for polymer, metal oxides and other materials of similar properties when needed.

Right pace:
The electrospinning device, which is powered by a high- voltage power supply, produces nano-fibers in the range of 10–800 nm in <u>diameter. Designed</u> for normal conditions of room temperature and low relative humidity, this device should be used in a ventilation <u>hood. It</u> may be used for polymer, metal oxides, and other materials when needed.

10.3   Current, Future, and Past Tenses

Many engineering reports use the *simple past* tense to describe the activities that already took place. However, the *present* tense should be used in the following cases (Alred et al. 2018):

1. Actions without time constraints
2. General truths
3. Routine activities (occurring in the past, present, and future)

4. Synopses of documents (to explain ideas presented in articles, books, videos, etc.)
5. Procedures being read (in current document)
6. Historical present (i.e., authors' opinions or contents in dated works)

Example 10.3 shows one sample sentence for each entry on the preceding list. To emphasize, you should use the present tense to describe the contents in a published work or the original author's opinion in that publication, although it was written in the past. This is so-called historical present, which emphasizes the content rather than the completion of action. You see this every day in newspaper headlines. For example, "AstraZeneca COVID-19 vaccines under investigation 'not shipped to Canada', officials say" *(Source: globalnews.ca/news/7695938/astrazeneca-blood-clots-vaccine/).*

The simple future tense is rarely used in academic writing for engineering publication because it expresses that something occurs after the present. It uses phrases like *is going to* or words like *will* before the main verb. The simple future tense can be used for proposal writing, indicating the plan of activities. Proposal writing is beyond the scope of this book, although a good part of this book can be used as guidelines for proposal preparation.

Furthermore, consistent use of tense is important to clarity. However, you can shift tense when a real change in time is needed. Example 10.4 compares illogical shift in tense with consistent use of tense. Imagine that you are describing a procedure for the assembly of a test apparatus; a shift from the past tone to the present one would confuse your readers.

Example 10.3. Cases for Present Tense

1)	This book, *Academic Writing for Engineering Publications,* <u>takes</u> its title from Tan's experience in engineering research and education.	*General truth*
2)	*Engineering* writing <u>is</u> unique, and it <u>is</u> different from that of business, fiction, materials, and physics.	*No time restriction*
3)	I arrive at the office 8 a.m. every day.	*Routine*
4)	This paper (Tan 2006) explains the basics and applications of informatics in public health, and the results demonstrate the need of further studies.	*Synopses of document*
5)	Readers of this book <u>read</u> Part II, *Organization of Ideas* first and <u>start</u> Part III, *Engineering Language Skills* later.	*Procedure being read*
6)	In his 1905 paper on "special relativity", Albert Einstein <u>argues</u> that space and time <u>are</u> bound up together.	*Historical presence*

Example 10.4. Consistence in Tense

Inconsistent: We <u>load</u> the reactants into the reactor before we <u>turned</u> on the heater.
Consistent: We <u>loaded</u> the reactants into the reactor before we <u>turned</u> on the heater.

You should avoid the unnecessary future tense in engineering academic writing. Example 10.5 uses the present tense for contents in current work, including cross-reference and contents in other sections. The reason is that your readers are reading *this* document when they hold it in their hands, although you wrote it before that moment.

Example 10.5. Avoiding Unnecessary Future Tense

Incorrect: The model <u>was</u> validated in Section 1; this section introduces the cases for further studies. The corresponding results <u>will be</u> presented in Section 3.

Correct: Section 1 <u>explains</u> the validation of the model, and Section 2 introduces the cases for further studies. The corresponding results <u>are presented</u> in Section 3.

Incorrect: <u>This</u> literature review <u>will focus</u> on the production of fuel products from carbon dioxide and water.

Correct: <u>This</u> literature review <u>focuses</u> on the production of fuel products from carbon dioxide and water.

10.4   Point of View

Point of view is another factor affecting the clarity of writing. Point of view can be expressed in first person (e.g., *I, we*), second person (e.g., *you*), or third person (e.g., *she, her, it*) pronouns. Using consistent point of view in the same sentences helps avoid confusing writing.

Impersonal point of view is normally used in academic writing for engineering publication. An impersonal point of view effectively emphasizes the subject matter, and personal point of view emphasizes the doer. In practice, it is occasionally necessary to use the pronoun in technical writing to avoid awkward sentences, where *one* or *the writers* instead of *I* or *we* is used.

Avoiding rapid shifting of point of view ensures a smooth flow of thoughts (*see* Example 10.6). An illogical shifting from the third person, which is typical in technical writing, to the first person in mid-sentence will likely confuse the readers.

Example 10.6. Point of View and Positive Tone

Awkward: "It is noted that the performance of Sample A is not as good as that of state-of-the-art MEAs reported by Gas." (Liu et al. 2009).

Revision: <u>We</u> noted that MEAs (reported by Gas) performed better than Sample *A* did. *[The revision improves conciseness and positiveness.]*

10.5   Positive and Negative Tones

Like verbal communication, written communication also shows your tones. Your tone of communication reveals your personality, as well as your general attitude toward the subject matter. A positive tone enhances your image. For this reason, you should avoid negative words such as *unfortunately, sadly, absurd*, and *ridiculous* in academic writing. Example 10.7 shows how to achieve positiveness in writing by transforming the negative into opportunities.

Example 10.7. Transform the Negative into Opportunity

Negative: <u>Sadly,</u> the conversion efficiencies of these reactors are still <u>too low</u> for them to become a solution for the industry.

Positive: ~~Sadly, t~~The conversion efficiencies of these reactors are still ~~too~~ low and <u>further research is needed</u> for them to become a solution for the industry. *[Implies opportunity]*

Additionally, readers might view the message as negative when they see the words *not, no*, and those with the same function (e.g., *neither/nor*). Thus, you should use them only when necessary. As seen in Example 10.8, you can transform a negative tone into a neutral tone by replacing *not* and the word modified by *not* with a prefixed word, phrase, or clause of the same meaning. For example, *impersonal* for *not personal* and *need improvement in accuracy* for *not accurate*).

Example 10.8. Using Prefixed Words to Reduce Negativity

Negative: <u>Personal</u> point of view is normally <u>not</u> used in academic writing for engineering publication.

Neutral: <u>Impersonal</u> point of view is normally used in academic writing for engineering publication.

Example 10.9. Positive and Negative Thoughts

Negative: Results show that 70% of the conclusions are misleading.

Positive: Results show that 30% of the conclusions are supported with data.

The next section introduces grammatical voices. Grammatical voice and grammatical tone are different, and they can be easily mixed up. With continual practices, however, you will understand their difference and use them properly.

10.6   Active and Passive Voices

A grammatical voice can be *active* or *passive*. Both active and passive voices are effective, depending on the emphasis. *See* Example 10.10. The active voice emphasizes the performer of action; the passive voice emphasizes the subject being acted upon. Make sure that the voices are consistent in the same sentences.

Both active and passive voices are effective, depending on the emphasis. The active voice emphasizes the performer of action; the passive voice emphasizes the subject being acted upon. Make sure that the voices are consistent in the same sentences.

Example 10.10. Active Voice and Passive Voice

Active: When driving a car, the engine pistons <u>turn</u> the crankshaft and <u>drive</u> the powertrain under the car. The tires <u>rotate</u> and <u>move</u> the car forward.

Active: <u>Check</u> gas pressure and temperature before opening the reactor.

Passive: Gas pressure and temperature <u>should be checked</u> before opening the reactor. *[In the passive voice, it is not clear whether the pressure and temperature are checked; the active voice clearly indicates the performer of the action.]*

Active voices are highly valued in English. Active voice is generally preferred because it is concise, it improves clarity, and it avoids confusion. In some cases, passive voice fails to identify the performer (*see* Examples 10.11 and 7.8). Therefore, you should use active voice when you can.

Example 10.11. Passive Voice Causing Vagueness

Passive: Hurrying to finish the data collection on time, two computers <u>were used</u> simultaneously. *[This sentence implies that two computers were hurrying!]*
Active: Hurrying to finish the data collection on time, <u>we used</u> two computers simultaneously *[It is clear in this sentence that the performer of the action is <u>we</u>, the authors.]*
Passive: The computers <u>were turned off</u> before the tests finished.
Active: The <u>lab manager turned off</u> the computers before the tests were finished.

Example 10.12. Passive Voice Causing Wordiness

Passive: Anti-EpCAM antibodies are anchored on the surfaces of microposts to capture the CTCs by ligand affinity.
Active: Anti-EpCAM antibodies anchored on the surfaces of microposts capture the CTCs by ligand affinity.

However, it does not mean that the passive voice is useless. The passive voice is effective or necessary when it is unnecessary to identify the performer of the action; it achieves emphasis over the object. For instance, the first sentence in Example 10.13 emphasizes the *abacus* and its value; the inventor or who used it was less important.

Example 10.13. Passive Voice Emphasizing Object

1. <u>The abacus</u> was early used for arithmetic tasks.
2. This paper <u>was</u> published in 1905.

Passive voice may be appropriate to the explanation of a process or a procedure, which is useful to describing methodology. This is important to engineering publication because, in many cases, it is not necessary to state who the performer is. The readers should be able to replicate the work following the procedure. Example 10.14 is written in passive voices to describe an experimental procedure (Dolan et al. 2010).

Remember that using the appropriate grammatical voice is important to the writing in English. Non-native English writers tend to follow the habit of their native languages. The passive voice is used frequently in some languages, but in others, not at all. For example, Japanese frequently uses the passive voice.

Example 10.14. Passive Voice for Description of Procedures

2. Experimental

Experiments <u>were performed</u> in a batch 69 mL reactor constructed from stainless steel 316 tubing (w17% Cr, w13 Ni, w3% Mb, w2% Mg, <0.03% S), which <u>was heated</u> in a muffle furnace preheated to the set point temperature. The reactor <u>was filled with</u> 5, 10, 15, 25, or 35 mL sodium carbonate solution containing 4% by weight microgranular cellulose (Sigma) and 0.04% by weight 5% Pt/Al_2O_3 (Sigma, powder). The 5, 10, 15, 25 or 35 mL slurries permitted study of headspace fractions of 93, 86, 78, 64 and 49% respectively.

Prior to heating, the reactor headspace <u>was filled and evacuated</u> with argon several times to purge the headspace of molecular oxygen. Finally, the headspace <u>was filled</u> with 15 psi of argon to allow for sufficient pressures for gas sampling in cases where little gas <u>was produced</u> …

10.7 Defining Terms Before Use

10.7.1 Defining Technical Terms

Definition is important to clarity and accuracy. Many graduate students assume that readers are in the same areas of research and have the necessary background. This is a typical mistake in engineering academic writing. As explained in Section 3.3, the readers vary from college students to experts in the field. Thus, using technical terms without definition first may cause misinterpretation. Defining terms for your readers is essential to clarity of your writing.

Example 10.15. Defining Terms Being Used

The slurry products were collected and filtered using Whatman No.1 filter paper to remove <u>residual solids</u>. These solids on the filter paper were then heated at 110 °C until its weight became constant. <u>The residual solid yield, defined as the mass ratio of residual solid to the volatile content of input cattle manure</u>, was then calculated (Yin 2011).

You can define terms by analogy, cause, component, equation, example, exploration of origin, and so on. Giving examples (introduced by *for example*, *for instance*, *such as*, and *like*) is the clearest approach. However, terms defined by their causes are especially effective in writing academic articles aimed at international readers. Definition by components helps the readers understand the concept with different smaller pieces (see Example 10.16). Under certain circumstances, however, exploration of origins may be more effective than others; it is especially useful to terms with unfamiliar Greek and Latin roots.

Example 10.16. Definition by Components

Formal definition:

Combustion is a chemical reaction between fuels and oxidants that produces energy in the form of heat and light and new chemical species. *[Source: Meyers, 2012, Encyclopedia of Sustainability Sci & Tech]*

Definition by component:
The chemical formula shows that combustion requires <u>fuel and oxygen</u>. <u>Oxygen</u> is normally taken from air, and <u>fuel</u> can be any flammable material. A combustion starts with <u>ignition</u>, and the <u>heat</u> produced sustains the combustion process. <u>Light</u> is generated at the same time. Furthermore, new <u>chemical species</u> are produced by combustion; they are mostly <u>air pollutants</u> or <u>solid wastes</u>.

On the other hand, you need to use definition techniques with care. Avoiding circular definition and definition with "is when," "is where," and the like help with both clarity and conciseness of writing. (*See* Examples 10.17 and 10.18.)

Example 10.17. Avoid circular Definition

Incorrect: <u>Spontaneous</u> human combustion is the combustion of a living human body <u>spontaneously</u>.

Correct: <u>Spontaneous</u> human combustion is the combustion of a living human body without an external source of ignition.

10.7.2 Consistent Naming of Same "Things"

The "Never use the same wthe other hand, you need to use definition ord twice" rule that you learned from high school does not apply to engineering academic writing. You should use the same word to describe the same *thing* in the same document. Otherwise, your writing will confuse many of your readers.

Example 10.18. Avoid "Is When" in Definition

Incorrect: A supercritical fluid <u>is when</u> its temperature and pressure are above the critical point.

Correct: A supercritical fluid <u>is a substance</u> at the temperature and pressure above the critical point.

Example 10.19 compares inconsistent and consistent use of terms in a rough draft (with nonessential text omitted for clarity). Numerous terms appear in these three paragraphs and one figure caption title. The rough draft lacks clarity because of the exchangeable use of *method* and *technique*, which are different terms. In addition, the rough draft omits *CTC enrichment*, which modifies the word *methods*, in many places. In contrast, consistent terms greatly enhance the clarity of writing

Example 10.19. Using Same Term for Same Thing

Inconsistent terms	Consistent terms
Figure 1 presents the <u>classification</u> of <u>CTC enrichment methods</u>. They include <u>single- and hybrid-modality methods</u>. The <u>single-modality methods</u> are further classified into <u>biophysical methods and biochemical methods</u> based on the dependent property of CTCs or blood cells. ... *Figure 1. <u>Classifications</u> of <u>CTC enrichment techniques</u>* Despite substantial progresses in <u>single-modality methods</u>, most of them have not been widely adopted in clinical utility.... A <u>hybrid method</u> combines two or more <u>single-modality techniques</u> For example, ..., 2) <u>multiple enrichments</u> are more likely to provide higher purity. We further divide these <u>hybrid-modality techniques</u> into <u>one- and multi-step method...</u>	Figure 1 presents the <u>classification</u> of <u>CTC enrichment methods</u>. They include <u>single- and hybrid-modality CTC enrichment methods</u>. The <u>single-modality CTC enrichment methods</u> are further classified into <u>biophysical and biochemical CTC enrichment methods</u> based on the dependent property of CTCs ... *Figure 1. <u>Classification</u> of <u>CTC enrichment methods</u>* Despite substantial progresses in <u>single-modality CTC enrichment methods</u>, most of them have not been widely adopted in clinical.... A <u>hybrid CTC enrichment method</u> combines two or more <u>single-modality CTC enrichment methods</u>. ... For example, ... 2) <u>multiple CTC enrichment methods</u> are more likely to provide higher purity. In this paper, we further divide these <u>hybrid-modality CTC enrichment methods</u> into <u>one- and multi-step methods</u> ...

10.8   Analogy and Simile

Analogies and similes are two typical figures of speech used in engineering writing. Analogies often use the structure *as ... as* for comparison. A simile directly compares two objects or concepts using the word *like* or *as*. Analogies normally are followed by necessary elaborations to enhance clarity. *See also* Example 10.20.

Example 10.20. Analogy and Simile

Analogy   Outlining in writing is <u>as important as</u> framing in building construction.
Simile   Outlining in writing is <u>like</u> the framing of a house; a constructor cannot build a functional house without a well-structured framing work. [*Elaboration*]

Properly used figures of speech improve the effectiveness of communication between the writer, who is typically an expert in the scope of writing, and the novice readers. To some extent, figures of speech also make writing interesting and lively.

Despite their effectiveness, you should use figures of speech only when it is necessary, especially in academic writing aimed at international readers. Non-native English speakers might translate figures of speech literally, causing a misunderstanding of your ideas.

10.9 Clarity

Clarity is *essential* to academic writing. Clarity requires both logical organization of ideas and skillful use of language. They form the main part of this book. Examples of techniques include coherent outlines, structured paragraphs, convincing arguments, sufficient details, smooth transitions, correct emphases, precise words, accurate punctuation, and much more.

As explained in Part II, organizing ideas in a logical sequence within an outline is critical to achieving clarity, coherence, and unity. Drafting converts each of the sentences in the outline into a paragraph. You ought to ensure that each sentence in the outline is converted into a paragraph with a single controlling idea supported by a body of evidence. This helps maintain your ideas organization. The rest of this book introduces language skills for clear and concise writing.

Chapter 11 introduces various **paragraphing** techniques. A typical paragraph in academic writing consists of three basic components, in the order of sequence:

1. The topic sentence
2. The body sentences
3. The concluding sentence

This paragraph structure ensures one single idea in each paragraph.

Each sentence plays an important role in the paragraph. The topic sentence is a statement of the controlling idea. The body sentences support the statement with evidence, examples, analyses, etc. The same paragraph ends with a concluding sentence, which reminds the readers of the paragraph's main point or hints the topic of the next paragraph.

Emphasis and subordination are two complementary techniques that differentiate the importance of phrases, clauses, and sentences. Without them, readers may have to guess their relative importance. In addition, appropriate transition is important to clarity and coherence; transition is also essential to the smooth flow from one paragraph to the next or from one sentence to another. See also Chaps. 11 and 12 for elaboration and examples.

Precise wording is another basic requirement of clear writing. Make sure you know the precise meaning of the word in its context when you choose synonyms. Often there is a subtle difference of a word in different contexts, which may be difficult for non-native English speakers to identify. The subtle difference can change the tone of your writing when using words of the same meaning. For example, Anglo-Saxon words are often short and simple, whereas Old French words are longer and sophisticated. The difference is how formal you want your writing to be, although they may mean the same thing.

Table 10.2 compares some of the Anglo-Saxon and Old French synonyms that appear in engineering academic publications (source: Wikipedia 2021). In a pair of words, one may be used more often than the other, but neither Anglo-Saxon nor Old French words dominate in academic writing. For example, the Anglo-Saxon word *ask* sounds neutral, whereas the Old French word *enquire* may sound overly formal; *reason* (Old French) is a better choice than *sake* (Anglo-Saxon) in technical writing. Nonetheless, both *end* and *finish* are acceptable.

It is essential to learn the precise meanings of words. In addition to reading this book, you are encouraged to read some handbooks for formal writing (e.g., Alred et al. 2018). For example, understanding *antonym* and *synonym* allows you to choose the right words. Antonyms are pairs of words with opposite meanings, such as *good/bad, interesting/boring, long/short, expensive/economical*); synonyms are words that have the same or similar meanings, for example, *average/mean, biannual/biennial, notorious/infamous*, etc.

Table 10.2 Some Anglo-Saxon and Old French synonyms

Anglo-Saxon	Old French	Anglo-Saxon	Old French
Answer	Reply	Help	Aid/assist
Ask	Enquire	Leave	Exit/depart
Bear	Carry	Live	Reside
Behaviour	Manner	Need	Requirement
Belongings	Property	Sake	Reason/cause
Brittle	Fragile	Smell	Odor
Buy	Purchase	Span	Distance
Deal	Amount	Thorough	Exhaustive
Deem	Consider	Tough	Difficult
End	Finish	Uphold	Support
Follow	Ensue	Weak	Feeble/faint
Fall	Autumn	Wild	Savage
Friendly	Amicable	Worthy	Valuable

Furthermore, precise wording requires continual growth in vocabulary. Reading, listening, and conversing in English can help grow your vocabulary for academic writing. International doctoral students, for example, are encouraged to read at least one journal article a day and as much as possible. The reading materials should be beyond technical; reading newspapers, magazines, biographies, etc. also helps you in building vocabulary. In addition, watching television after work and listening to radios while driving is useful to vocabulary development too. (See 10.2. Writing with Simple Language; 13.2.1. Phrasal Verb; 13. Phrases and Word).

Avoiding vague words certainly improves the clarity of academic writing. Your fellow professionals expect your clear and accurate communication. For example, words like *important*, *good*, *well*, *bad*, and *thing* are subjective with multiple meanings and interpretations. Instead, use concrete words and choose the right ones. *See* Chapter 12. Sentence for the techniques of proper use of modifiers and intensifiers.

Example 10.21. Subjective Modifiers (Good/Poor/Well)

Vague:

- All sharp diffraction peaks indicate <u>good</u> crystallinity of the sample.
- This is because $Na_2WO_4 \cdot 2H_2O$ has a <u>poor</u> solubility in DMF.
- Our results agree <u>very well</u> with theirs.

Clear:

- Sharp diffraction peaks indicate <u>strong</u> crystallinity of the sample.
- This is because $Na_2WO_4 \cdot 2H_2O$ has a <u>low</u> solubility in DMF.
- The difference between our results and theirs <u>is less than 5%.</u>

10.10   Transition and Connection

10.10.1   Transition Between Paragraphs and Sections

Transition between paragraphs is accomplished with the same techniques used between sentences, except that the transitional elements may be different and longer. The transitional elements can be one of the following:

1. A transitional paragraph (*see* Examples 10.22 and 10.23)
2. A sentence summarizing the preceding paragraph
3. A question, at the end of the preceding paragraph, to be answered at the beginning of the new paragraph

---

Example 10.22. A Transitional Paragraph Between *Model Description* **and** *Results*

With the model developed, we first studied the water-substrate interfacial behavior, followed by the effects of water layer thickness on the collision behavior of a nanoparticle on a pristine silver substrate. Finally, the effects of surface energy of the substrate on the collision behavior are reported. The next section presents all these results. (Khodabakhshi et al. 2021).

Example 10.23. A Transitional Paragraph Ending *Introduction*

The rest of this article is organized as follows. Section 2 focuses on biophysical and biochemical single-modality methods, which are the key elements of hybrid modality methods. Then Section 3 presents hybrid method as one- and multi-step methods according to the enrichment steps involved. Different combinations are introduced in this section with explanations of coupling mechanism. Finally, conclusions and future perspectives are presented in Section 4.

10.10.2   Connecting Sentences in Paragraph

A well-structured paragraph is characterized with *clarity*, *unity*, *coherence*, and *adequate development*. *Unity* refers to the extend that the topic sentence embodies a single idea or concept: one idea in each paragraph. *Coherence* is the connection of sentences into a logical single point of view by transitional words.

Appendix A1 is an incomplete list of transitional words and phrases. Using the right transitional words (*additionally, alternatively, similarly, in addition, in contrast, specifically, furthermore, however*, etc.) is key to the coherence of the paragraph.

Time-order words can improve clarity and coherence by connecting sentences. They can also be combined with lists using numbers or bullet points. Table 10.3 lists some time-order words. You can combine these words to achieve lively and clear writing.

These transitional words usually introduce paragraphs, sentences, or items in a list. In addition, easy-to-follow instructions can be described with simple short steps, which can be organized with time-order words. Simple enumeration in the order of sequence (*first, second, third*, and so on) also provides effective connection between sentences within the paragraph.

Table 10.3 Time-order words used in technical writing

Before	Begin	In between	Last
Earlier	To begin	Next	Finally
In the past	First	Then	In conclusion
Preceding that	To start with	Later	In the end
Previously	At the beginning	Consequently	At the end
Prior to	To begin	After	At last
Before	Initially	Subsequently	Ultimately

Examples 10.14 and 10.25 compare the effectiveness with and without transitional words connecting sentences. Both show the power of time-order words. Using them appropriately allows you to write with clarity and effectively communicate your ideas with your readers.

Example 10.24. Using Time-Order Words for Coherence

Published:
PVA nanofibers were made with different applied voltages, tip to collector distances and deposition times. The morphologies of these electrospun filters were then characterized by SEM images coupled with an automated image analysis method. Using NaCl airborne nanoparticles in the size range of 10–125 nm, the single-layer and multilayer filters were also evaluated in terms of filter quality factor. The effects of the electrospinning parameters on filter quality factor were determined to identify the important factors affecting filtration performance of PVA nanofibrous filters.

Revised:
PVA nanofibers were <u>first</u> made with different applied voltages, tip to collector distances, and deposition times. <u>Then,</u> the morphologies of the electrospun filters were characterized by SEM images coupled with an automated image analysis method. <u>Later,</u> the single-layer and multilayer filters were also evaluated in terms of filter quality factor using NaCl airborne nanoparticles in the size range of 10–125 nm. The effects of the electrospinning parameters on filter quality factor were <u>finally</u> determined to identify the important factors affecting filtration performance of the PVA nanofibrous filters.

Example 10.25. Description of Sequence with Time-Order Words

Published:
A custom-made electrospinning setup was used in this study for filter sample preparation. The relative humidity and temperature of the air inside the housing were $39 \pm 4\%$ and 23 ± 3 °C, respectively. A 5-ml syringe was loaded with a solution of PVA polymer, which has a molecular weight of 89,000–98,000 g·mol-1 (Sigma Aldrich Canada). The desired solution concentration of 10% w/w was prepared by diluting the PVA in distilled water at 90 °C and stirring overnight (Appendix B). A 22-gauge stainless steel capillary needle with an inside diameter of 0.413 mm was attached to the syringe. The syringe was mounted on a syringe pump (Kd Scientific), which was used to control the flow rate to 0.3 ml·hr^{-1}. A lab jack was used to adjust the vertical distance between the capillary needle and the grounded collector. A high-voltage power supply (ES50P-5 W) was employed to apply the high voltage between the capillary needle and an aluminum collector.

Revision option 1, restructuring the sentences:
A custom-made electrospinning device was used in this study for filter sample preparation; the relative humidity and temperature of the air inside the housing were 39 ± 4% and 23 ± 3 °C, respectively. Prior to electrospinning, a solution of PVA polymer (with a concentration of 10% w/w) was prepared by diluting the PVA in distilled water at 90 °C and stirring overnight (see Appendix B). PVA has a molecular weight of 89,000–98,000 g·mol^{-1} (Sigma Aldrich Canada). To begin, the solution was loaded into a 5-ml syringe with a 22-gauge stainless-steel capillary needle, which has an inside diameter of 0.413 mm. After that, the syringe was mounted on a syringe pump (Kd Scientific), which was used to control the flow rate to 0.3 ml·hr^{-1}. Subsequently, A lab jack was used to adjust the vertical distance between the capillary needle and the grounded collector. Finally, a high-voltage power supply (ES50P-5 W) was used to apply the high voltage between the capillary needle and an aluminum collector.

Revision option 2, using a list:
A custom-made electrospinning device was used in this study for filter sample preparation; the relative humidity and temperature of the air inside the housing were 39 ± 4% and 23 ± 3 °C, respectively. Samples were prepared following these steps.

- Prior to electrospinning, a solution of PVA polymer (with a concentration of 10% w/w) was prepared by diluting the PVA in distilled water at 90 °C and stirring overnight. PVA has a molecular weight of 89,000–98,000 g·mol^{-1} (Sigma Aldrich).
- To begin, the solution was loaded into a 5-ml syringe with a 22-gauge stainless-steel capillary needle, which has an inside diameter of 0.413 mm.
- After that, the syringe was mounted on a syringe pump (Kd Scientific), which was used to control the flow rate to 0.3 ml/hr.
- Subsequently, A lab jack was used to adjust the vertical distance between the capillary needle and the grounded collector.
- Finally, a high-voltage power supply (ES50P-5 W) was used to apply the high voltage between the capillary needle and an aluminum collector.

10.11   Conciseness

Conciseness is important to engineering writing. You can achieve conciseness in writing by simple sentence structure, verb-based writing, concise wording, short phrases, and so on. A frequently used strategy is the removal of redundant words. Nonetheless, the removal should not sacrifice clarity, coherence, or unity. Additionally, you should avoid overuse of descriptors; otherwise the writing becomes stuffy and wordy.

Example 10.26. Concise Writing
Wordy: This indicates an effective photon utilization, which is an important factor in a photocatalytic process.
Concise: This indicates an effective photon utilization, which is important to photocatalysis.

The rest of this book details the techniques to achieve clarity and conciseness in writing. However, it takes continual effort to master these skills. You can refer to this book when you write.

10.12 Practice Problems

Question 1: Revise these sentences with the correct tenses.

1. Tests of the filters by separating cancer cells from diluted blood sample (blood: PBS = 1:6) were carried out to evaluate the filters performance. Compared with pristine filter, the functionalized filter can improve the recovery rate from 80% to 95% with less than 0.1% WBCs contamination.
2. Combination of the biochemical method of LBM and biophysical method of HBM can utilize the high flow rate of HBM and compensate for the low specificity of HBM by LBM. For instance, Liu et al. [92] combined DLD and immunoaffinity-based fishbone microchannel as one platform. Fig. 9a shows the platform configuration. The first enrichment unit consisted eight parallel DLDs, the outlets of which were then connected in series with another DLD. When the blood sample flowed through the first unit, CTCs and some larger WBCs were deflected gradually to the center of DLD (with cut-off diameter of 5–6 µm). Then the CTCs and some larger WBCs from the center outlet of DLD flowed into the second enrichment unit, which had two antibody-grafted microchannels in series. As a result, CTCs were retained on the channel substrate by antibodies while WBCs were flowed away. The DLDs in both parallel and series configurations raised the processing rate to 9.6 mL/min. By this platform, recovery rates of ~90% with 17–52% purity were realized for enriching MCF-7 cells from diluted whole blood sample.

Question 2 (the voices): Rewrite these sentences by changing passive voices to active voices.

3. In the first enrichment step, WBCs labeled by magnetic beads were depleted by negative IMS. Then the second enrichment step was carried out on using a microchannel for filtration.
4. In the second enrichment unit, IMS was used to separate CTCs from WBCs.
5. After that, WBC-beads complexes were separated from CTCs by IMS.
6. Tests of the filters by separating cancer cells from diluted blood sample (blood: PBS = 1:6) were carried out to evaluate the filters performance. Compared with pristine filter, the functionalized filter can improve the recovery rate from 80% to 95% with less than 0.1% WBCs contamination.
7. It has been introduced in Section 2.2.2 that the reduction of MnO_2 nanospheres by oxalic acid can achieve cell release [27].
8. An optimal height-to-diameter ratio of ~about 13 with diameter of ~230 nm was found. Capture efficiencies of 85–95% were obtained to isolate 10–100 MCF-7 cells from 1 mL whole blood, and the captured cells showed intimate interactions with nanopillars.
9. Test results showed that a release efficiency of 96.2% can be obtained with 90% viability.
10. Under the influence of magnetic field, the labeled CTCs experienced lateral migration while unaffected blood cells were swept by buffer flow.
11. It has been introduced in Section 2.2.1 that the AuNPs were fabricated for photothermal therapy.
12. The ecological risk characterization will be done per the methodologies and recommendations presented in the U.S. EPA Screening Level Ecological Risk Assessment Protocol. Finally, in task three, a stochastic model will be employed to evaluate risk in the presence of variability and uncertainty and to alleviate the issue of not knowing how conservative the estimate is. The Monte

Carlo method will be used to simulate the stochastic processes in this research. Monte Carlo analyses have been used to account for the wide variability of exposures in many studies.

13. Molecular dynamics (MD) simulations are performed to analyze the effects of condensed water layer on the collisional dynamics of nanoparticles on a substrate. The collision of 5-nm silver particles on a dry or wet silver substrate is studied. The coefficient of restitution, which is inversely proportional to the stickiness of nanoparticle to the surface, is used as the main characteristic to describe the collision. (Khodabakhshi et al. 2021)

14. The surface energy of the substrate is controlled by varying its Lennard-Jones potential parameters that is used to describe the particle-water and particle-substrate interaction. (Khodabakhshi et al. 2021)

15. Such peculiar behavior can be reasonably explained by the mechanisms of energy dissipation during collision including plastic deformation, particle-substrate interaction, and particle-water interaction. It is shown that all the dissipation mechanisms depend on both water layer thickness and surface energy of the substrate. (Khodabakhshi et al. 2021)

16. Detailed descriptions on physics and chemical processes during the collision between moving nanoparticles and the substrate is still very poorly understood on a molecular level, although many experimental and theoretical studies have been performed in the continuum flow regime and with particles of larger than a few micrometers. (Khodabakhshi et al. 2021)

17. Some thermal energy of colliding nanoparticles, which is transformable into the translational kinetic energy, should be taken into account in addition to the critical velocity. (Khodabakhshi et al. 2021)

18. It was revealed that the relative hardness of the particle relative to that of the substrate affects energy dissipation and *CoR*. (Khodabakhshi et al. 2021)

19. Despite a relatively large body of literature on nanoparticle and substrate collision, an interesting parameter, relative humidity (RH) or water content, has not been widely investigated. When most of existing researches on nanoparticles have been focused on dry collision, a study showed the adhesion of microparticles to a surface is promoted by increasing the moisture content of ambient air. This effect has been attributed to the capillary force of the condensed water layer on the surface. The controversial roles of the humidity in influencing the nanoparticle and substrate collision needs to be illustrated based on molecular level analytical investigations. (Khodabakhshi et al. 2021)

20. To address the role of humidity, the water-surface model can be added by integrating the parameters which were obtained empirically and able to predict the physical properties of liquid water. Recently, nanoparticles and substrates made of silver were used as a model system to study the interfacial phenomena. The embedded-atom method[17] (EAM) has been extensively used to model the silver-silver potential interaction between the atoms of a silver particle. As reported by Foiles et al., a consistent set of embedding functions and pair interactions were determined empirically by fitting to the physical properties of the pure silver, to describe the FCC silver particle or surface. To model the silver-silver potential interaction between atoms of two separate particles, a Lennard-Jones (LJ) interaction has been often employed. (Khodabakhshi et al. 2021)

21. The collision of a nanoparticle on a wet substrate is investigated using molecular dynamics. It was shown that the results of MD simulations to calculate the coefficient of restitution in a collision process are in good agreement with the experimental results. (Khodabakhshi et al. 2021)

22. The initial representation of the simulation box is presented in Figure 1. The substrate and the nanoparticle are both cut out from a grid of FCC structure. A thin layer of water, with the initial thickness of L_w, is placed on the substrate. (Khodabakhshi et al. 2021)

23. Only the collision of a single nanoparticle on a substrate is modeled and the interactions between the particle and the surrounding air molecules are not considered. Silver is chosen as the particle

and substrate material. The interaction between the atoms of the particle or substrate is modeled by an embedded atom model (EAM). Based on this model, the potential energy of atom i is calculated based as follows. (Khodabakhshi et al. 2021)

24. The material parameters for silver reported by Foiles et al. are used for the modeling of silver atoms. For the interaction between particle-substrate, particle-water, and water-substrate atoms, the standard (6–12) LJ relationship is applied. Same as our previous work, water molecules are modeled using the modified TIP3P model. (Khodabakhshi et al. 2021).

25. The SHAKE algorithm is used to fix the internal geometry of the water molecules. The Lenard-Jones pair interactions and the partial charges for all the atom types in the simulation box are summarized in Table 1. To make weakly adhesive collisions, the interaction strength of the nanoparticle is controlled by reducing its LJ parameter. Surface energy of the substrate is also controlled by changing its LJ parameter, ε. (Khodabakhshi et al. 2021)

26. The reduction of the interaction strength of the nanoparticle and substrate is based on the assumption of a partial oxide layer created on their surface weakening the surface interaction. (Khodabakhshi et al. 2021)

27. The particle-particle particle-mesh (PPPM) method is used to compute the long-range electrostatic interactions with a cut-off distance equal to 12 Å and a root-mean-square accuracy of 10^{-4} (Khodabakhshi et al. 2021)

28. All the simulations were started from the Maxwell-Boltzmann velocity distribution for all atoms at 300 K. The number of atomic layers is chosen to ensure that the energy from the collision is absorbed by the substrate and propagated through it properly. Two outermost atomic layers of the substrate were fixed to prevent the movement of the substrate while the other layers were left to vibrate normally. The NVT ensemble was used initially with a Nose-Hoover thermostat, which maintained the system temperature at 300 K. (Khodabakhshi et al. 2021)

29. During collision the last two bottom layers were kept fixed in time and the substrate temperature was controlled with Langevin dynamics applied to the next eight atomic layers. All other atoms were followed with a Newtonian dynamics. (Khodabakhshi et al. 2021)

30. A same double-layer structure of water molecules near a graphene surface was observed in a previous work. (Khodabakhshi et al. 2021)

31. These characteristic peak distributions of the water molecules are maintained even as the number of water molecules increases. (Khodabakhshi et al. 2021)

32. The double-layer structure of water molecules on the substrate could be justified by the concept of surface energy of the substrate. (Khodabakhshi et al. 2021)

33. In what follows, at first the water-substrate interfacial behavior is studied. Afterward, the effects of water layer thickness on the collision behavior of a nanoparticle on a pristine silver substrate is studied. Finally, the effects of surface energy of the substrate on the collision behavior is investigated. (Khodabakhshi et al. 2021)

34. It was reported by Akaishi et al. that the arrangement of hydrogen bonds in the first layer is in a way that there exist some dangling bonds toward the outside of this layer. However, since the distance of the second layer is higher than that of the first layer, and the LJ interaction is short range, the attraction applied to the water molecules in the second layer is reduced. The LJ interaction decreases considerably for $z > 8$ Å, and as a result, the water molecules outside the second layer are weakly affected by the substrate. Moreover, it was reported that the H-bonds in the second layer are all saturated with the H-bonds in the first layer and therefore, there is no dangling H-bond toward the outside of the second layer. As a result, the water molecules outside the second layer are arranged with no specific arrangement and no clear peaks are found beyond the second layer. Since the short-range LJ is the only interaction, just the water molecules close to the substrate are affected. (Khodabakhshi et al. 2021)

35. It should be noted that the increment of the distance between the particle and the substrate is around 1.5 Å. (Khodabakhshi et al. 2021)

36. It should also be noted that the particle-substrate interaction increases with increasing the impact velocity. (Khodabakhshi et al. 2021)

37. In Figure 7a, the radius of gyration is plotted as a function of water layer thickness for all the studied impact velocities. (Khodabakhshi et al. 2021)

38. It is indicated that energy loss in colliding nanoparticles in the plastic collision range is seemingly tied to the degree of permanent deformation and to the corresponding deformation modes. The type of plastic deformation regime is determined by the collision velocity. The nanoparticle gets squeezed in the direction parallel to the collision axis and elongated in the directions perpendicular to the collision direction. Consequently, the nanoparticle is deformed like a pancake. (Khodabakhshi et al. 2021)

39. The coefficient of restitution as a function of the water layer thickness (L) for the collision of a 5-nm particle on a silver substrate is depicted in Figure 4a. This kind of bimodal behavior was also reported previously in the study of collision of nanoparticles on a dry substrate. (Khodabakhshi et al. 2021)

40. So far, the collision of a 5-nm particle on an ideal silver substrate was investigated under dry and wet conditions. Moreover, the surface energy of a substrate could be controlled by some functional groups on the outer surface of the substrate. In the preceding sections, it was shown that the surface energy of the substrate controls the particle-substrate interaction strength as well as the water behavior near the substrate. (Khodabakhshi et al. 2021)

41. In Figure 7, the coefficient of restitution is plotted as a function of the water layer thickness over the whole range of studied velocities. (Khodabakhshi et al. 2021)

42. Plastic deformation is studied by plotting the radius of gyration of the nanoparticle in Figure 8.

43. The average particle-substrate interaction is plotted in Figure 9a. Similar to Section 3.1, increasing the distance between the particle and the substrate can be explained by the water density distribution near the substrate. (Khodabakhshi et al. 2021)

44. In a humid environment, water molecules in the air are condensed and form a water layer on the substrate. (Khodabakhshi et al. 2021)

45. First, just the results for a dry collision are discussed, then the effects of humidity are considered in the next paragraph. (Khodabakhshi et al. 2021)

46. Although the particle-substrate interaction is reduced, the particle-water interaction compensates for this reduction. (Khodabakhshi et al. 2021)

47. In summary, it was shown that the collision properties of a nanoparticle to a substrate depends on relative humidity. While for high impact speeds, the adhesion of the particle to the substrate with humidity can be reduced. Since the collision properties of nanoparticles to the substrate is determined by the particle impact energy dissipation, the mechanisms of energy dissipation in dry and wet impact were compared with each other. (Khodabakhshi et al. 2021)

48. Great efforts have been carried out experimentally to develop battery materials that can improve LIB performance. (Li and Tan. 2020)

49. Numerical modeling and simulation of LIB has been developed over the past decades and been used by many researchers as an alternative tool to optimize battery design parameters. (Li and Tan. 2020)

50. In most engineering situations, it is the average velocity, pressure, etc. that are of interest, and the details of all the turbulent eddies are not required. (Bai et al. 2013)

51. The field variables at the previous time step n are assumed to be known, and are not reflected in eq. (19). (Bai et al. 2013)

52. Though most wind tunnel test results could not be obtained, it is reasonable to anticipate that the values using three-dimensional CFD would be more accurate than those using two-dimensional ones and that three-dimensional CFD simulations have important practical significance for blunt bodies. (Bai et al. 2013)

53. To identify the flutter derivatives of the three sections for zero angle of attack, forced motion simulations were conducted using the driving signal amplitudes of eq. (29). (Bai et al. 2013)

54. It can be summarized from Table. 1 that the capture efficiency of this method is higher than 40% for blood sample.

55. The immunomagnetic separation can be performed either in tubes (or wells) or by microchannels.

56. *[Part of data analysis]* Peaks below an absolute intensity of 1000 were deleted to remove noise, which corresponds to a relative peak intensity of about 0.1%. Higher intensity noise peaks still present in the spectra were distinguished from real peaks by assessing the variability of the accurate mass value of all the peaks in the mass spectrum over the 1-minute infusion. For a real peak the variability of the measured m/z for a compound is less than ±2 ppm whilst noise peaks show a much higher variability (> ±2 ppm). To remove noise peaks each spectrum was split in half, and the first 30 seconds and the second 30 seconds were averaged and molecular formula assigned separately within an error of ±2 ppm, according to the process described below. Only formula assignments found in both 30-second sections are kept. It is assumed that the extraction process mixes the extracted compounds in the solvent droplet sufficiently to ensure that the sample composition remains constant for each one-minute infusion. (Fuller et al. 2012)

57. The recent development of commercially available surface mass spectrometry ionization techniques such as liquid extraction surface analysis (LESA) and desorption electrospray ionization (DESI) enables analysis of a wide range of organic compounds with high spatial and therefore high time resolution from RDI samples (for details see the Experimental Section). LESA has been used previously for the analysis of biological samples and pesticides, but applications for environmental samples have not yet been described in the literature. Another online extraction technique, nano-DESI, was recently presented by Roach et al. and was applied to atmospheric aerosol filter samples. Importantly, LESA and (nano-) DESI require no off-line sample preparation, such as solvent extraction and solvent evaporation. Reducing the number of sample preparation steps also reduces the possibility of introducing artifacts. (Fuller et al. 2012)

58. The first task aims to tackle the roadway geometry characterization limitation between the models and provide an accurate method in characterizing road sources in the air dispersion models. … Many of the MSATs deposit in the environment, where humans are exposed via indirect pathways, contributing further to the risk already caused by inhalation exposure. Moreover, since there are only a few air quality standards for the MSATs, in the second task, we will use the cancer unit risk estimates and noncancer inhalation reference values to estimate risks from MSATs (and HAPs). This investigation will be done per the methodologies and science contained in the U.S. EPA Human Health Risk Assessment Protocol (HHRAP). The ecological risk characterization will be done per the methodologies and recommendations presented in the U.S. EPA Screening Level Ecological Risk Assessment Protocol (U.S. EPA 1999). Finally, in task three, a stochastic model will be employed to evaluate risk in the presence of variability and uncertainty and to alleviate the issue of not knowing how conservative the estimate is. The Monte Carlo method will be used to simulate the stochastic processes in this research. Monte Carlo analyses have been used to account for the wide variability of exposures in many studies. *[Curtsey: M. Munshed]*

Question 3 (positive tones): Convert the following sentences from negative to positive or neutral tone.

59. Unfortunately, due to the generally low peak intensities of the identified species, MS/MS analysis for further structural identification was not possible. (Fuller et al. 2012)
60. This correlation is not as strong if all compounds up to m/z 350 are considered. (Fuller et al. 2012)
61. Finally, an investigation with different forced vibration amplitudes shows that amplitude effects on the flutter derivatives of bluff bodies are not significant if the amplitudes are not large enough. (Bai et al. 2013)
62. However, the flutter derivatives A and B are not presented below because wind tunnel results are not available for them. (Bai et al. 2013)
63. There are not obvious differences between them, though the maximum amplitudes that Noda et al. [30] used were reached. (Bai et al. 2013)
64. However, Yasuna et al. (1995) found that this trend was not so obvious for Geldart B particles unless the solids volume fraction was close to the maximum packing limit. (Du et al. 2006)
65. All the simulated voidage profiles show that in the spout, the voidage decreases with the increasing bed height, illustrating a qualitative agreement with the experimental tendency except in the higher part of the column, where the maximum voidage does not appear at the axis. (Du et al. 2006)
66. Unfortunately, by digging out in literature, we found that a detailed examination to the effect of drag models on CFD modeling seems unavailable. (Du et al. 2006)

Question 4 (transition and connection): Rewrite these sentences with proper transitions and connections:

67. It can be seen that the present three-dimensional CFD simulations mostly give better results than the DVM method and ANSYS workbench, though sometimes all results are in good agreement. For A, which is well-known to be a critical parameter for flutter [29], the present three-dimensional CFD method has obtained exact results when compared to wind tunnel results. It can be concluded that no matter if the structure belongs to streamline body or bluff body, the present three-dimensional CFD method gives better predictions for flutter derivatives of bluff bodies, which have relatively poor aerodynamic stabilities than the DVM method and ANSYS workbench. (Bai et al. 2013)
68. In spouted bed systems, the volume fraction of particles can vary from almost zero to the maximum packing limit, leading to much more complex behavior of drag forces than that in normal fluidization systems. By incorporating various drag models into the two-fluid model, the present study is conducted with the aim of fully understanding the influence of the choice of drag models on simulation and thereby laying a basis for the CFD modeling of spouted beds. (Du et al. 2006)
69. A possible explanation for this diminished advantage of using high porosity separator for 38120 LIB is that the heat generation for 38120 cylindrical battery is much higher than coin cell battery, eventually leading to the enhancement of mass transfer and electrochemical reaction for both batteries with high and low porosity separators (see Eq. 5, 6 and 15). Wang et al. [33] monitored the temperatures of a CR2032 type coin cell LIB, the results showed that the temperature rise on the coin cell surface was always below 0.05 °C during battery operation at various C rate. In this study, if we set the battery modeling at a constant temperature of 298.15 K, an obviously higher energy density for using 95% porosity separator (84.1 Wh/kg) than 35% porosity separator (81.8 Wh/Kg) was thus observed at high discharge rate of 10 C. (Li and Tan 2020)

70. Figure 4 shows the electrolyte concentration distribution across a single battery cell with different separator thicknesses and porosities. The electrolyte concentration decreased from anode to cathode at the end of discharge, forming a gradient across battery cell. When increasing the separator thickness from 5 to 100 μm, the differences between electrolyte concentration gradients within the separator domain were not significant. While, the concentration gradient in the separator domain is decreased with the increasing separator porosity from 35% to 95%. (Li and Tan 2020)

71. The excessive heat accumulation and uneven temperature distribution of LIB may degrade the LIB performance [28]. As shown in Table 2, the heat conductivities of the separator and the electrolyte are much less than those of other LIB components. As a result, the thermal resistance of a 20-μm separator counts for about 70% of the total resistance in the battery along the radial direction. Thus, it is important to improve the thermal properties of battery separator. (Li and Tan 2020)

72. The primary standards are set to provide public health protection, and the secondary standards provide public welfare protection. The current NAAQS includes 12 different standards for 7 pollutants: carbon monoxide (CO), lead (Pb), nitrogen dioxide (NO_2), ozone (O_3), particulate matter 10 micrometers or less in aerodynamic diameter (PM_{10}), particulate matter 2.5 micrometers or less in aerodynamic diameter ($PM_{2.5}$), and sulfur dioxide (SO_2). Table 1 shows the nitrogen dioxide (NO_2) national air quality standard ("Primary National Ambient Air Quality Standards (NAAQS) for Nitrogen Dioxide I U.S. EPA", 2020); the new 1-hour NO_2 NAAQS was promulgated in the United States in February 2010. *[Curtsey: M. Munshed]*

73. The standard considers the multi-year average of the 98th percentile of the yearly distribution of 1-hour daily maximum nitrogen dioxide concentrations. The 100 ppb (188 μg/m^3) standard was set to protect against the respiratory effects (airway inflammation and asthma exacerbations) of nitrogen dioxide (Guarnieri & Balmes, 2014). The primary sources of nitrogen dioxide emissions are vehicles and coal powerplants (Howard, Thé and Soria, 2019). *[Curtsey: M. Munshed]*

Question 5 (conciseness): Revise the following writing to improve conciseness.

74. The objective of this study is to explore the links between the size-resolved filtration efficiency of nanofibrous media mounted in portable air cleaners in a chamber test with results from the media mounted in a cone-shape filter holder in a more conventional duct test. (Givehchi et al. 2021)

75. We collected available field measurements of personal exposure or infiltration factors of ambient air pollutants to compare with modeled results. *[Source: Hu et al. Environment International (2020) 144: 106018]*

76. Uncertainty in flow rate was calculated based on the larger of the manufacturer's reported uncertainty of ±5% or the standard deviations from three replicate tests.

77. On the other hand, it has been shown that the aerosol type (KCl or diesel particles in size range of 30–1000 nm) significantly affect CADR.

78. Model results under various conditions show the separator thickness had a strong impact on the battery energy density.

79. It is timely to examine the current status of SO_2 scrubbing technologies. *[Source: doi.org/10.108 0/10473289.2001.10464387]*

80. There are many substances in the air which may impair the health of plants and animals (including humans). *[Source: newworldencyclopedia.org/entry/Air_pollution]*

81. Although a number of studies have been done both experimentally and theoretically on the effect of loading on filtration efficiency of micron-sized fibers, limited studies were found for this effect on nano-sized fibrous filters.

82. Besides particle loading, charge masking or discharge of filters over time may also cause reduction in the filtration efficiency.

83. IPA causes the filtration efficiency of media M2 to be negative for sub-500 nm particles in the chamber test... Overall, IPA caused the filtration efficiency of tested electrospun nanofibrous media to be decreased.

Question 6:   Transition and connection.

1. Identify at least one transitional paragraph between the sections in this sample paper.
2. Identify at least ten conjunction words in the sample paper.
3. Identify five wordy writing errors in the sample paper, and suggest revisions to improve conciseness.

- **Sample paper:** Wang Y, Li H. 2020. Bio-chemo-electro-mechanical modelling of the rapid movement of *Mimosa pudica*. *Bioelectrochemistry* 134: 107533. (sciencedirect.com/science/journal/15675394)

Paragraphs

11.1 Statement-Evidence-Conclusion Structure

A paragraph is composed of multiple sentences that are related to the single controlling idea of the paragraph. Thus, the sentences in the paragraph must be organized coherently. All paragraphs share certain common structural characteristics, and this chapter introduces the techniques for writing well-organized paragraphs.

A normal paragraph is composed of three key elements: topic sentence, body sentences, and concluding sentence. Thus, you should aim for four to five sentences per paragraph.

The topic sentence is a statement that prescribes the controlling idea of the paragraph. Following the topic sentence is the body of evidence that supports the statement. The body elaborates on the statement using logical argument, examples, analyses, etc. The paragraph ends with a concluding sentence, which reminds the readers of the main point of the paragraph and signals the idea of the next paragraph. Each of the sentences in the paragraph plays an important role in communication between you and your readers.

The *statement-evidence-conclusion* structure is typical to writing in English, but not necessary for works produced in another language. Therefore, it takes some practices to master the paragraphing techniques.

11.1.1 Topic Sentence

The topic sentence declares the single controlling idea of the entire paragraph. The topic sentence unifies the controlling idea of the paragraph and guides the writing with the order of the sentences to follow. It is important to put the topic sentence at the beginning, typically the first sentence, of the paragraph because most readers look to the topic sentence to determine the single idea of that paragraph.

Occasionally, the first sentence is not a topic sentence. First, a topic sentence is optional to a paragraph that continues developing the same idea as that in the previous paragraph. Second, a transitional sentence may proceed the topic sentence, which links the current paragraph to the previous one.

Examples 11.1 and 11.2 are acceptable paragraph structures. The contents are from the same article written by Fuller et al. (2012) unless stated otherwise. Example 11.2 is the same paragraph as in Example 4.46. The paragraph begins with a topic sentence (*There are a number of analytical–chemi-*

Z. Tan, *Academic Writing for Engineering Publications*,
https://doi.org/10.1007/978-3-030-99364-1_11

cal techniques) followed by two examples to support this idea. One example is *(AMS) techniques* and the other, *aerosol samples collected on filters or impactors*. The closing sentence is omitted, and this type of short structure is acceptable despite its imperfection.

Example 11.1. Paragraph with Statement Followed by Evidence
Figure 4 shows the O/C and N/C ratios of the most intense peaks in the mass spectra as a function of their molecular mass from both stages. For clarity only the most intensive 200 peaks in the mass spectra are shown here, which correspond to about 80% of the total ion intensity. The molecular formulas from all 20 extraction points were combined and their intensities from each point summed.

Example 11.2. Paragraph with Statement and Supporting Evidence

There are a number of analytical–chemical techniques that allow measuring the composition of aerosol particles with high time resolution. Online aerosol mass spectrometry (AMS) techniques, for example, allow for very highly time-resolved particle composition studies. However, such measurements are demanding with respect to manpower and other resources and are usually performed only for a few weeks at a specific site. Thus, such measurements rarely provide insight into long-term trends of particle composition. In contrast, aerosol samples collected on filters or impactors over long time periods are more readily available but have mostly a rather low time resolution of the order of a day or more, and their chemical analysis usually involves time-consuming.

In contrast, Examples 11.3, 11.4, 11.5, and 11.6 illustrate problematic paragraphs with various errors. Typical errors include multiple ideas in one paragraph and the statement of the controlling idea at the end of paragraph.

Example 11.3 shows part of a manuscript that a graduate student asked me to review. First, the paragraph needs a topic sentence at the beginning. The writer put the current scope (*high speed train*), scopes of earlier works (*most studies... and some further...*), and other research methods (*The major methods to study a train*) into one single paragraph. The rapid pace of writing causes idea clutter. The revisions, following the *statement-evidence-conclusion* paragraph structure, help improve coherence in writing.

Example 11.3. Multiple Ideas in One Paragraph

[Begins Introduction] In the last few decades high speed train (with speeds up to x kph) have been designed and built in a number of countries. Coupled with this increase in speed is a general tendency for the weight of the train to reduce. Hence, the risk of overturning potentially increases, especially when a train travels during periods of bad weather such as strong cross winds, rainstorms or sandstorms, etc. Most of current studies were concerned with train under cross wind, including aerodynamic characteristics and running stability with various factors such as configurations of vehicles and infrastructures, the yaw angle and gust shape, etc. And some further address the primary cause of the aerodynamic forces such as the flow field. The

major methods to study a train under cross wind contains full-scale tests, wind tunnel tests, and numerical simulation. Panel methods and Multi-body simulation method were also applied in train's stability analysis. However, research on high-speed train traveling under severe bad weather such as thunderstorms (wind-rain condition), sandstorm, etc. are few.

Suggested revisions:
In the last few decades high speed trains (with speeds up to x kph) have been designed and built in several countries. For example, Germany built 100 from 1990 to 2012 with 10 more in the pipeline [citation]. There are even more in China with a total number of 200 between 2000 and 2010 [citation]… High speed trains become part of the model mobility system.

Coupled with this increase in speed is a general tendency for the weight reduction of high-speed trains to reduce. The weight of today's typical vehicle is only 20% of that 20 years ago. In 1990, a typical vehicle weighted about 1000 tons and now it is only 200 tons [citations]. Weight reduction is important to train speed manufacturing and operation.

This reduced weight comes with ~~Hence,~~ the increased risk of overturning ~~potentially increases~~. *[add evidence to support the statement.]* It is especially true when a train travels ~~during periods of~~ in bad weather such as strong cross winds, rainstorms, or sandstorms, ~~etc.~~

There have been a number of studies in high-speed train under challenging conditions. Most of them ~~current studies~~ were concerned with trains under cross wind. Factors of concern include aerodynamic characteristics and running stability with various ~~factors such as~~ the configurations of vehicles and infrastructures, the yaw angle, and the gust shape, etc. ~~And some further~~ Others address the primary cause of the aerodynamic forces such as the flow field. The major methods to study a train under cross wind ~~contains~~ include full-scale tests, wind tunnel tests, and numerical simulation. Panel methods and multi-body simulation method were also applied ~~into~~ train's stability analysis.

To the best of our knowledge, however, there is little research on high-speed train traveling under severe weather including thunderstorms, wind-rain condition, sandstorm, ~~etc.~~ *[add more evidence]*

Comments:
The suggested revisions divide an overly long paragraph into multiple short ones. Each new paragraph has one single controlling idea with the added text (underlined) or evidence to be added. (Refinement in language is beyond the scope of this example.)

Example 11.4. Statement at End of Paragraph (1)
"All 16 cities included in CAPES are involved in our study, namely Anshan, Beijing, Fuzhou, Guangzhou, Hangzhou, Hong Kong, Lanzhou, Shanghai, Shenyang, Suzhou, Taiyuan, Tangshan, Tianjin, Urumqi, Wuhan and Xi'an. These cities are geographically widely spread throughout China and vary greatly in climate. Their locations are shown in the Figure 1." (Zhou et al. 2013)

Suggested revisions:
Figure 1 shows the locations of the cities in this study. All 16 cities included in CAPES are involved in ~~our~~ this study, namely Anshan, Beijing, Fuzhou, Guangzhou, Hangzhou, Hong Kong, Lanzhou, Shanghai, Shenyang, Suzhou, Taiyuan, Tangshan, Tianjin, Urumqi, Wuhan, and Xi'an. These cities are geographically widely spread throughout China and they vary greatly in climate.

Example 11.5. Statement at End of Paragraph (2)

"Three long-span bridge deck cross sections are taken as examples in the present paper. They are named as sections G1–G3. Their geometrical features are shown in Figure 1, in which B is the chord length. Every model has two degrees of freedom, namely vertical translation h and rotation about its center. Section G1 can be treated as a streamlined structure, but G2 and G3 are typical bluff bodies with sharp edges. Particularly, the section G3, which has infamous aerodynamic instability because it is the prototype of the Tacoma Narrows Bridge in the USA, was destroyed …by a steady wind with the small velocity of 20 m/s. <u>So, it is necessary to investigate these structures.</u>" *(Source:* Bai et al. 2013*)*

Suggested revision: Relocate the last sentence to the beginning.

Example 11.6. Statement at End of Paragraph (3)

The photocatalysis begins when solar energy meets the TiO_2 semiconductor, exciting an electron of a TiO_2 molecule from the filled valence band to the empty conduction band. The energy required to excite the electron from the valence band to the conduction band is known as the band gap. When electrons migrate to the surface of the TiO_2 material and are trapped at the edge of the conduction band there, they serve as reduction centers, donating electrons to other acceptors. Similarly, the holes left in the valence band serve as oxidizing sites. <u>Figure 1 provides a schematic diagram of the photocatalysis process.</u>

Suggest revisions:
Relocating the last sentence to the beginning of the paragraph should bring the controlling idea up front. Note that this is a paragraph in a graduate-level course project report submitted by a native English speaker. It further emphasizes the difference between written and oral communication skills.

11.1.2 Paragraph Length

A paragraph should provide the readers with manageable subdivisions of thought. Otherwise, the writing appears to be disorganized. A series of short, undeveloped paragraphs break a single idea into several pieces, whereas a paragraph that is too long loses its conciseness and coherence. For this reason, it is generally not acceptable to write a single-sentence paragraph.

Note

A one-sentence paragraph is rare, but it is acceptable for a transition between two paragraphs or an entry of a list (Alred et al. 2018).

Example 11.7. Single-Sentence Paragraph

"While mass spectrometry is widely used to identify organic aerosol content, the combination with LESA and RDI sampling allows a much higher temporal sampling resolution than would be possible with other off-line techniques allowing one to identify marker compounds and to follow atmospheric processes in detail." (Fuller et al. 2012)

11.2 Lists in Paragraph

Appropriate use of a *list* is often more effective than a long paragraph for the presentation of certain information, for example, materials used, parts needed, steps in experiments, and key points for conclusions and recommendations.

The list entries may be words, phrases, clauses, and simple sentences. They can be listed vertically using numbers, letters, bullet points, or their combinations. The entire list maintains parallel structure and format throughout. All listed items normally begin with capital letters; listed sentences should also end with period marks.

A list can be a numbered (or lettered) list or use bullet points. Numbered lists save readers' time by focusing on steps of sequence, and a combination of numbers and letters allows you to list with subdivisions. Bullets are for items without rank or sequence. Nonetheless, you may still want to list the items in a logical order that supports your purpose of writing.

You should avoid overuse of lists. An unnecessarily long list with too many entries loses its effectiveness. In addition, make sure that you provide the context of the list. It would be difficult for your readers to follow your ideas if the paragraphs consist of entirely lists.

The following techniques are useful to creating effective lists:

- Maintain parallel structure for the entire list.
- List only comparable items of balanced significance.
- Capitalize the first word in each entry.
- Avoid commas or semicolons at the end of each item.
- Use ending punctuation (e.g., period) at the end of each item if the list comprises complete sentences.
- Avoid *and* and *or* at the end of any entry.

11.3 Equations in Paragraph

Equations are essential elements in engineering academic writing. You often need chemical, physical, and mathematical equations to further improve clarity of ideas. Regardless of the type of equation, each one takes at least one line by itself. Thus, you should avoid inserting equations into sentences. In addition, engineering writing avoids punctuation around equations, although some believe it helps with the flow of reading.

Regardless of format, equations are written with symbols and Arabic numbers – it is a mistake to write words or phrases in the equations. The symbols can be defined and listed in the *List of Symbols*. Alternatively, you can define the symbols and list them immediately after the equation where they are first used. In the latter case, the definition starts with an uncapitalized *where*. Furthermore, you need to italicize symbols, including the equations and the body text.

Example 11.8. Avoid Inserting Equation into Sentence

Incorrect: The radius of the nanoparticles is $r_p = 25$ Å.

Correct: The radius of the nanoparticles (r_p) is 25 Å.

Incorrect: Substituting Eqs 34 and 35 into Eq 19 gives:

$$\frac{4d_p\rho_p}{6}y_{(d_p)} = b\frac{\pi(d_p)^3}{6}\rho_p(n_0)^{\frac{2}{3}}\eta_{0(t=0)}, \text{then}$$

$$y_{d_p} = \left(\frac{b}{4}\right)n_0^{\frac{2}{3}}\eta_{0(t=0)}\pi(d_p)^2\left(1-e^{-kt}\right) \tag{36}$$

Correct: Substituting Eqs. 34 and 35 into Eq. 19 gives

$$\frac{4d_p\rho_p}{6}y_{(d_p)} = b\frac{\pi(d_p)^3}{6}\rho_p(n_0)^{\frac{2}{3}}\eta_{0(t=0)} \tag{36}$$

Dividing both sides of Eq. 36 by $4d_p\rho_p/6$ gives

$$y_{d_p} = \left(\frac{b}{4}\right)n_0^{\frac{2}{3}}\eta_{0(t=0)}\pi(d_p)^2\left(1-e^{-kt}\right) \tag{37}$$

Example 11.9. Chemical and Mathematical Equations

Incorrect: Hydrogen gas + oxygen gas → steam $\hfill (5\text{-}1)$

Revised: $2H_2(g) + O_2(g) \rightarrow 2H_2O(g)$ $\hfill (5\text{-}1)$

where g in parentheses stands for *gas*.

Incorrect: Uncertainty $= \sqrt{\text{Error fraction} \times \text{concentration} + (0.5 \times MDL)^2}$ $\hfill (4.3)$

Revised: $\alpha = \sqrt{(fC)^2 + (0.5MDL)^2}$ $\hfill (4.3)$

where α is the uncertainty; f is the error fraction; C is the concentration; MDL is the detection limit.

See Example 11.9. *See also* Sect. 16.2.18.

Number the equations consecutively throughout the final document. Most writers enclose the equation numbers with parentheses or brackets, which are on the right side of the equations and flush to the right margin.

In the body text, the equations are cross-referenced by equation numbers preceded with the word *Equation* or the abbreviation *Eq.* Either one is acceptable, but it should be used consistently throughout the document. However, do not list equations in the text unless the equation itself is part of the conclusions.

A long equation that cannot fit into one single line should be broken into multiple lines. The new lines should start with an equal sign or an operation sign ($+-\times\div$) that is *not* enclosed with parentheses or brackets. Align the left side of the first line and right side of the last line with other single-line equations on the same page. The lines in between can be centered or aligned by the right of equations. The long equation is numbered as one equation.

11.4 Practice Problems

Question 1 Revise the following paragraphs by supporting statement with evidence.

1. *[Rough draft]* As a label-free CTC enrichment method, the DEP can actively control the enrichment process by manipulating the electric field and achieve single-cell isolation [52, 57]. However, the motion of cell can be diverted by the dipole-dipole interactions between cells and the hydrodynamic forces from blood, causing the DEP response being disturbed. Therefore, the DEP-based method works with low loading capacity and low flow rate to reduce dipole-dipole interactions and hydrodynamic forces respectively [40]. Its processing rate is typically lower than 1 mL/h based on Table 1.
2. *[Rough draft; an end paragraph in one section in a review paper]* The hydrodynamics-based microfluidics is characterized with many advantages including label free, high processing rate (typically >10 mL/h), simple setup and operation, and free of cell release process (which facilitates purification or cell analysis). On the other hand, pre-processing dilution (which is about 10 times) is required to reduce the cell-cell interactions and non-Newtonian effects of blood. In addition, the size overlap and aggregation of cells impact the enrichment performance since the hydrodynamic force is determined by cell size.

Question 2 Following the guide in this book, analyze and rewrite the paragraphs in the *Introduction* of the sample paper.

- **Sample Paper:** C. Wu, J. Li, J. Liu (2020): Experimental study of a shallow foundation on spatially variable soils, *Georisk: Assessment and Management of Risk for Engineered Systems and Geohazards*, https://doi.org/10.1080/17499518.2020.1806333

Question 3 This is the second last paragraph of draft *Introduction* (Ge et al. 2015). Find the mistakes in the structure of this paragraph:

It is well known that DC streamer discharge has many advantages such as high energy efficiency, discharge stability as well as high efficiency in the generation of chemical active species. More importantly, O_3 and NO_2 production in this discharge mode are significantly less than that of in the first two discharge modes. Hence, the combination of streamer discharge and MnO_2 catalyst should be a promising method for indoor air VOCs removal. Until now, only a few studies concerned on the feasibility of applying streamer discharge combined with MnO_2 for the removal of indoor air VOCs.

Question 4: (Single Idea in One Paragraph) A student wrote the following paragraphs to review the current state of the research in CTC enrichment by filtration. The last paragraph serves as a summary:

1. How many ideas are presented in the last paragraph?
2. Did the student provide evidence to support the statements?
3. If you were the supervisor of the student, how would you recommend him to rewrite?

2.1.1 Filtration-based enrichment methods

A filtration-based technology isolates CTCs from blood because CTCs are larger, less deformable, and stiffer than healthy blood cells. The filtration unit can either a membrane filter or a microchannel. In filtration, CTCs are captured and retained while blood cells with smaller size flow through the membrane or microchannel.

Membrane filters for CTC filtration are fabricated by track-etching or micromachining (materials of parylene, silicon, metal, etc.). Their typical pore sizes are in the range of 6.5–8 μm. For example, Lin et al. reported a transparent parylene membrane filter with 8-μm circular pores. The CTC capture efficiency was 92% achieved in 2 min for 7.5 mL whole blood spiked with 90 cancer cells. The transparent membrane enables direct on-membrane characterization of the CTCs captured. Similarly, Kim et al. reported a membrane filter with an average pore size of 6.5 μm × 6.5 μm and the gap between adjacent pores was 6 μm. The capture purities of were 4.5–12% for filtration of 3 mL whole blood containing 30–100 cancer cells. In addition, Yagi et al. developed an automated CTC filtration system, where the membranes have 8 μm × 100 μm rectangular pores. In a clinical study, CTCs were detected in 40 out of 50 patients with lung cancer.

Different from membranes, microchannels enable precise manipulation of fluid behavior. CTCs are separated from the blood cells they flow through the gaps between pillars or weirs inside the microchannels [38]. For example, Qin et al. reported the microchannels with a series of weirs to isolate CTCs by controlling the aperture size between the weirs and the channel surface. The weirs with adjustable aperture sizes enable multiple filtration processes. Processing rate was increased by 128 parallel microchannels in one device. In a clinical study, the device captured more than 5 CTCs in 18 out of 22 mCRPC patients.

Filtration-based enrichment is characterized with low cost and simple operation without labeling process. A transparent membrane enables direct observation of the captured cells, and cells can be easily released by applying a reverse flow. However, some of the WBCs are also retained because of the occasional size overlap between CTCs (down to 10 μm) and WBCs (up to 20 μm). Moreover, the potential damage of CTCs and the clogging of filters should be considered in filter design and processing rate control.

Sentences 12

12.1 Sentence Construction

Sentences are constructed with words, phrases, clauses, and other elements following writing rules and structural styles. The *subject-verb-object* pattern is the most common sentence structure in English. Every sentence, except commands, must have a subject and a verb. A sentence using active voice should have a subject. In addition, the verb and its subject should be in grammatical agreement. Expletives (e.g., *there is*; *it is*) can be used to move the subjects away from their normal positions in the sentences, but it often leads to wordiness.

A sentence structure can be *simple, compound,* or *complex.* An independent clause forms a simple sentence; multiple independent clauses form a compound sentence. A compound sentence uses commas, semicolons, or coordinating conjunctions for clarity, and a complex sentence contains a subordinate clause or clauses. (*See* Sect. 12.2.)

Both long and short sentences are needed in academic writing provided that their meanings are clear to the readers. Uncomplex sentences are normally used in technical writing for the presentation of complex ideas. Simple sentences are easy to follow, and short sentences are generally used for direct statements. However, complex sentences are useful to explanations and convincing arguments. On the other hand, overly complex sentences likely confuse the readers whose native language is not English.

Experienced writers use various sentences in different lengths, structures, and complexities to achieve liveliness. A series of sentences of the same style become tedious and monotonous. Consequently, it loses the attention of the readers. Furthermore, varying sentence styles help create effective contrast and emphasis.

© The Author(s), under exclusive license to Springer Nature Switzerland AG 2022
Z. Tan, *Academic Writing for Engineering Publications*,
https://doi.org/10.1007/978-3-030-99364-1_12

Example 12.1 Sentence Structure Types

Simple:

- A pandemic is here.
- I am writing a book.

Compound:

- Aerosol samples can be collected using air filters for morphology analysis, but it is not the only way to characterize aerosol particles.
- The sizes of nanoparticles are 1in the range of 1–100 nm; various chemicals are present in the particles at different proportions.

Complex:

- The power to the corona charger will shut off automatically *[independent clause]* <u>when</u> the door is open *[dependent clause]*.

Compound-complex sentence:

- Fiber size is important to the filtration efficiency of a filter *[independent clause]*<u>, and</u> the filtration efficiency drops *[independent clause]* <u>when</u> the fiber size increases *[dependent clause]*.

12.2   Conjunction

A conjunction connects words, phrases, or clauses in a sentence to explain the relationship between these elements. Table 12.1 compares four types of conjunctions: coordinating conjunction, correlative conjunction, subordinating conjunction, and conjunctive adverb.

Table 12.1 Conjunctions

Conjunction	Example	Function
Conjunctive adverb	accordingly, alternatively, besides, consequently, for example, for instance, in another word, further, however, moreover, nevertheless, on the other hand, similarly, then, therefore, thus, etc.	Connect two independent clauses
Coordinating conjunction	and, but, for, nor, or, so, yet, etc.	Connect two sentence elements with identical functions
Correlative conjunction	either/or, neither/nor, not only/but also, both/and, whether/or, not/but, etc.	Used in pairs of equal importance
Subordinating conjunction	after, although, as, as if, because, before, despite, even if, even though, if, in order that, only if, rather than, provided (that), since, so that, that, though, unless, until, when, where, whether, while, etc.	Connects the elements of a sentence and distinguishes their relative importance

A coordinating conjunction joins two sentence elements with identical functions, and correlative conjunctions correlate and work in pairs to join words, phrases, or clauses. Both coordinating and correlative conjunctions indicate equal importance. They also introduce elements in parallel structures as explained in the preceding section. Therefore, the elements (words, phrases, or clauses) in both parts of the pairs should have similarity and follow the same grammatical form. (*See* Sect. 12.3)

Subordination effectively distinguishes the relative importance of two sentence elements. The main idea of the sentence depends on the position of the subordinating element, and the more important one normally precedes the less important one. For example, both sentences in Example 12.2 are logical, but their main ideas differ from each other.

Example 12.2 Subordinating Conjunction Using *although*

- <u>Although</u> cobalt based solvents are effective in absorbing NOx, they create secondary pollution. *[The main idea is "effective cobalt based solvents."]*
- Cobalt based solvents create secondary environmental pollution, <u>although</u> they are effective in absorbing nitric oxide. *[The focus of this sentence is cobalt based solvents create secondary environmental pollution.]*

Example 12.3 Subordinating Conjunction Using *because*

Providing excuses <u>Because</u> the relative humidity level in the laboratory is not controlled properly and the dehumidifier in the testing compartment failed, the results contain artifacts.

Providing facts The results contain artifacts <u>because</u> the relative humidity level in the laboratory is not controlled properly and the dehumidifier in the testing compartment failed.

12.3 Parallel Structure

Parallel structures and conjunctions are commonly used for the construction of effective sentences. They are used to achieve clarity, conciseness, emphasis, and variety with less words, phrases, or clauses. Parallel structures also allow readers to anticipate the meaning of elements in the sentences or paragraphs.

Example 12.4 Parallel Structures in Writing
Parallel words:

- No pain, no gain.
- Corona viruses carry <u>either DNA or RNA</u>, never both.
- A sophisticated air quality monitor may be used to detect these air pollutants: <u>sulfur dioxide</u>, <u>nitric oxides</u>, <u>ozone</u>, and <u>particulate matter</u>.

Parallel phrases:

- <u>Both</u> medical professionals <u>and</u> patients are exposed to the airborne viruses.
- The reaction requires a high pressure, a low temperature, <u>and</u> oxygen free condition.

Parallel clauses:

- <u>Either we work</u> together to fight the pandemic, <u>or we let</u> the pandemic beat us all.
- Your technical proposal <u>not only must</u> have a list of objectives, <u>but also should</u> include a list of deliverables.
- In typical scholarly writing, the authors are expected <u>that they would start with</u> background information, <u>that they would elaborate on</u> the methodology, <u>and that they would present</u> the results to the readers.

An effective parallel structure can be constructed by repeating an element of the sentence. The element can be an article, pronouns, prepositions, and so forth. The sentence that you just read is parallel structured. Parallel structures can be constructed using words, phrases, clauses, and sentences, but they should not be mixed. To emphasize, a parallel structure that begins with a word, phrase, or clause must continue using words, phrases, or clauses. (*See also* Example 12.4.) Moreover, parallel structures are developed with sentences having the same grammatical structure: adjectives are paralleled by adjectives, nouns by nouns, verbs by verbs, etc.

In summary, consistence is key to effective parallel structure because the readers can anticipate the properties of the elements to come. On the contrary, inconsistence can confuse your readers and weaken your argument. (*See* Example 12.5.)

Example 12.5 Consistence vs. Inconsistence Parallels	
Wrong	This research is <u>interesting and a challenge</u>.
Right	This research is <u>interesting and challenging.</u>
Unparallel	A sophisticated air quality monitor may be used to detect these air pollutants: <u>sulfur dioxide, volatile organic compounds, ozone</u>, and *measuring* <u>particles</u>.
Parallel	A sophisticated air quality monitor may be used to detect these air pollutants: <u>sulfur dioxide, volatile organic compounds, ozone</u>, and ~~measuring~~ particles.
Unparallel	Wildfires are more <u>intense</u>, droughts are <u>commonplace</u>, and in humid climates hurricanes are more and more expected.
Parallel	Wildfires are more intense, droughts are more common~~place~~, and hurricanes are more expected than before in humid climates.
Unparallel	In typical scholarly writing, the authors are expected <u>that they would start</u> with background information, <u>that they would elaborate</u> on the methodology, <u>and that results would be presented</u> to the readers. *[Shifting from voice breaks the parallel structure.]*
Parallel	In typical scholarly writing, the authors are expected <u>that they would start</u> with background information, <u>that they would elaborate</u> on the methodology, <u>and that they would present</u> the results to the readers.

12.4   Sentence Variety

Monotonous writing may be boring to your readers and yourself too. You can write interesting sentences with modifiers (*see* Sect. 12.7.2). However, overuse of modifiers can be monotonous too. The best approach is to use a variety of sentences. There are several techniques to achieve sentence variety, including varying sentence length and inverting word order. They are briefly introduced here.

12.4.1 Varying Sentence Length

You can vary sentence lengths by connecting short independent clauses with subordinations (*see* Table 12.1). However, using too many instances of "and" in one sentence is a typical mistake that non-native English writers make. You can also combine short sentences into a long one by converting verbs into adjectives. However, verb-based writing enhances clarity and conciseness.

12.4.2 Varying Word Order

Example 12.7 shows that either inverting word order or inserting an element enhances sentence variety. This technique is effective in achieving emphasis, providing details, and controlling pace.

Example 12.6 Sentence Variety (Length and Tone)
Monotone The distillation column is 20 m tall, <u>and</u> its diameter is 4 m, <u>and</u> its surface roughness is 5 μm.
Revised The distillation column<u>, which is</u> 20 m tall and 4 m in diameter, has a surface roughness of 5 μm.

Example 12.7 Sentence Variety (Word Order)

Inverting word order:

- **Normal:** The photo of a blackhole <u>has never been</u> so clear before.
- **Inverted:** <u>Never has</u> the photo of a blackhole <u>been</u> so clear before.

Inserting an element:

- A high-speed air flow creates friction on the surfaces, <u>both top and bottom,</u> of the plate. *[This insertion emphasizes location.]*

12.5   Grammatical Agreement

Grammatical agreement is a form of correspondence among subject, verb, object, and other elements of a sentence. For instance, the subject, verb, and number in the same sentence must be in grammatical agreement; pronouns must agree in *person*, *number*, and *gender*. *See* Sect. 2.9. Neutral Language for gender agreement.

12.5.1 Subject-Verb Agreement

Pay attention to singular and plural pronouns. *One*, *each*, *series*, and *portion* are singular nouns; occasionally they precede plural nouns. Indefinite pronouns (*all*, *more*, *none*, and *some*), either in their singular or plural forms, can be used with mass nouns or count nouns, respectively. Relative pronouns (*that*, *which, who, whom, whose*) can precede either singular verbs or plural verbs, depending on the context.

Example 12.8 Subject-Verb Agreement (1)

- One out of ten truck drivers interviewed at the rest areas is female. *[The verb must agree with one instead of truck drivers or rest areas.]*
- Some of the carbon dioxide has become supercritical fluid. [*Carbon dioxide* is a mass noun.]
- Most of the truck drivers do not know that air in the truck cabin is polluted *["drivers" is plural]*.
- A large portion of journal articles in this field is devoted to the relationship between climate change and the solar radiation.
- The reading of the pressure gage, which is recorded by the computer, is 20 kPa.
- Pay attention to the temperatures of the samples that are monitored every minute.

Not all words ending with -*s* are plural subjects. For example, *analysis* is singular, and its plural form is *analyses*. Abstract nouns (*ethics, mathematics, news, physics*, etc.) appear to be plural but are singular. Non-native writers in English may want to pay close attention to the following confusing cases:

1. A verb must agree with its subject regardless of the number of the subjective complements.
2. Subjects that express measurement, weight, mass, or the like take singular verbs.
3. Some words like *scissors* are plural only.

Example 12.9 Subject-Verb Agreement (2)

Wrong   The topic of this report are nanomaterials.
Right   The topic of this report is nanomaterials. *[The word nanomaterials may be distracting.]*

Example 12.10 Subject-Verb Agreement (3)

Wrong   At least three numbers are needed to calculate the standard deviation of those numbers.
Right   At least three numbers *is* needed to calculate the standard deviation of those numbers.

Example 12.11 Subject-Verb Agreement (4)

- A <u>pair</u> of scissors *was* used to remove the string.
- <u>Scissors</u> *were* used to remove the string.

Regardless if singular or plural, the title of a long document, including book and thesis, takes a singular verb in a sentence.

Example 12.12 Subject-Verb Agreement (5)

Wrong　*Air Pollution and Greenhouse Gases* <u>are</u> published by Springer.
Right　*Air Pollution and Greenhouse Gases* <u>is</u> published by Springer.

12.5.2 Compound Subject-Verb Agreement

Compound subjects are joined by a conjunction (*see* Sect. 12.2). Two parts joined by *and* normally take a plural verb (Example 12.13). However, there are two exceptions to this grammatical rule:
1. When the compound subject is generally thought to be a single unit. *See* Example 12.14.
2. When a compound subject is modified by *each* or *every*, it becomes singular. *See* Example 12.15.

Example 12.13 Compound Subject-Verb Agreement (1)

Wrong　<u>The severity and frequency</u> of these events <u>has</u> introduced the conversation of global warming and climate change to the forefront of many political debates.
Right　<u>The severity and frequency</u> of these events <u>have</u> introduced the conversation of global warming and climate change to the forefront of many political debates.

Example 12.14 Compound Subject-Verb Agreement (2)

Wrong　Peanut butter and jelly <u>are</u> my favorite.
Right　Peanut butter and jelly <u>is</u> my favorite.
Wrong　Two and three <u>equal</u> five.
Right　Two and three <u>equals</u> five.

Example 12.15 Compound Subject-Verb Agreement (3)

Wrong　<u>Each</u> conclusion and recommendation <u>are</u> important to the readers.
Right　<u>Each</u> conclusion and recommendation <u>is</u> important to the readers.
Wrong　<u>Each</u> figure and table <u>have to</u> be numbered.
Right　<u>Each</u> figure and table <u>has</u> to be numbered.

A compound subject can be made up of one singular and one plural element that are joined by *or* or *nor*. This type of subjects requires the verb to agree with the closer element.

Example 12.16 Compound Subject-Verb Agreement (4)

Wrong   (Either) Table 1 or <u>Figures 4-6 supports</u> the argument.
Right   (Either) Table 1 or <u>Figures 4-6 support</u> the argument.
Right   (Either) Figures 4-6 or <u>Table 1 supports</u> the argument.

Note

You can split the compound subject and use two sentences with simple subjects when you are unsure whether the verb is plural or singular.

12.6   Fragmented Sentences

A sentence becomes fragments when it omits key sentence elements, such as subject or verb. A stand-alone subordinate clause or phrase is also a sentence fragment. Sentence fragments can occur in not only long complex sentences but also simple ones. They often result from informal spoken English.

- **Missing Verbs.**
  Missing verbs cause fragmented sentences. A sentence must have at least one verb. It is not a problem when the sentence is short and simple, but missing a verb is. It may occur unintentionally, especially when the sentences become complex. In addition, verbals cannot replace verbs for the same grammatical functions.

Example 12.17 Sentence Fragment (Missing Verbs)

Fragment   And record the reading displayed on the screen.
Sentence   He records the reading displayed on the screen.
Fragment   Supporting your argument with examples.
Sentence   You need to support your argument with examples.

- **Missing Subjects.**
  Missing subjects often occur when the writers assume them to be something by default, confusing the readers. *See* Example 12.18.

Example 12.18 Sentence Fragments (Missing Subjects) (1)

Fragment   The increasing carbon dioxide concentration in the atmosphere causes to rise over the last century.
Sentence   The increasing carbon dioxide concentration in the atmosphere causes the ocean temperature to rise over the last century.

- **Other Sentence Fragments.**

Sentence fragments often result from the use of relative pronouns or subordinating conjunctions. Using explanatory phrases (*such as, for example*) may also create sentence fragments. Example 12.19 compares a sentence fragment with a correct sentence.

Example 12.19 Sentence Fragments (Missing Subjects) (2)

Fragment   The biological approach comes with some benefits. <u>For example, contributes</u> to environmental sustainability.

Sentence   The biological approach comes with some benefits. <u>For example, it contributes</u> to environmental sustainability.

12.7 Clauses

12.7.1 Adjective Clauses

Adjective clauses are modifiers following nouns. They can be particularly tricky for non-native writers of English because clauses are formed in a variety of ways in different languages. *See* Sect. 12.9.

Example 12.20 Adjective Clauses

Wrong   The one who <u>she</u> is giving the presentation is my student.
Right   The one who ~~she~~ is giving the presentation is my student.
Wrong   The student is giving a presentation <u>who is standing in front of the screen</u>.
Right   The student <u>who is standing in front of the screen</u> is giving a presentation. [*Comment*: The adjective clause *who is standing in front of the screen* modifies *student instead of presentation*. Therefore, it must immediately follow the word *student*.]

12.7.2 Dependent and Independent Clauses

A clause can be dependent or independent. An independent clause stands alone like a simple sentence. Otherwise, it is a *dependent* clause or subordinate clause. Each sentence must contain one or more independent clauses, but dependent clauses are used for effective expression of thoughts and establishment of their relative importance. See also Sect. 12.9.2.

Example 12.21 Clause Importance

Equal importance   The landfill site in Waterloo region is in the City of Waterloo. It is used by the cities of Cambridge, Kitchener, and Waterloo.

Subordinated   The landfill site in Waterloo region, <u>which is in the City of Waterloo</u> *[dependent]*, is used by the cities of Cambridge, Kitchener, and Waterloo.

12.8   Emphasis

Emphasis is essential to clarity as it stresses the primary ideas of sentences and paragraphs. Effective emphasis can be accomplished by alternating position, using repetition, using intensifiers, varying sentence length, changing sentence type, and so forth. Using special fonts like *italics,* **bold** face, underlining, and CAPITAL letters also enhances emphasis. However, you should only use them occasionally to avoid visual clutter. Another common approach is to use phrases like *again, most importantly,* and *foremost.* You can also use dash marks for similar purposes (*see* Punctuation).

12.8.1 Emphasis by Position

Although the *subject-verb-object* pattern is most familiar to English writers, you occasionally may construct an inverted sentence to emphasize your main ideas. You can catch the readers' attention by placing the elements in an unusual order to achieve emphasis. Broadly speaking, you should start a sentence, a paragraph, or a document with the ideas that you wish to emphasize. The reason is that the beginning and end elements of a sentence, a paragraph, or a document catch readers' eyes.

Example 12.22 Inverted Sentence

Normal   Uncomplex sentences are preferred in technical writing for the presentation of complex ideas. *[Emphasizes the value of uncomplex sentences]*

Inverted   In technical writing, uncomplex sentences are preferred for the presentation of complex ideas. *[Emphasizes technical writing; it may not be true for another type of writing.]*

Preceding the main idea of a sentence with less important elements is a typical mistake that non-native speakers in English frequently make. Non-native English writers often put the purpose, location, reason, and so on before the main idea of a sentence. There may be a cultural reason behind this error. For example, most Chinese people consider being straightforward as being impolite; they respond to an inquiry beginning with reasons even in their verbal communications. These indirect introductory elements, however, will demote the importance of the main idea and confuse the readers.

Example 12.23 Emphasis by Positions of Elements

Incorrect   Below headspace, the volume is filled with solvent, catalyst, and biomass.

Correct   The volume below headspace is filled with solvent, catalyst, and biomass.

Incorrect   In order to establish a quantitative relationship between winter road condition and vehicular air emissions, multiple linear regression models were developed. [*The scope of the thesis work is models.*]

Correct   Multiple linear regression models were developed to establish a quantitative relationship between winter road condition and vehicular air emissions

Incorrect   If the process can be improved, solar fuels would provide many benefits for society.

Correct   Solar fuels would provide many benefits to society if the process can be improved.

Incorrect   Another factor affecting the clarity of writing is point of view.

Correct   Point of view is another factor affecting the clarity of writing.

Another way to emphasize by position is using lists. You can list your ideas allowing the readers to understand their order of importance in sequence. See Sect. 11.2.

12.8.2 Emphasis by Sentence Length

Strategically varying a sentence length achieves emphasis in a paragraph. A short sentence that follows a long one, for example, usually catches the readers' attention.

Example 12.24 Emphasis by a Short Sentence

We have conducted a start-of-the-art literature review on the methodology and key findings of nanofiber for filtration. All lead to a conclusion that, regardless of the advances that were made recent years in nanomaterials, nanofiber by electrospinning is a relatively new field of research. More research is needed.

12.8.3 Emphasis by Repetition

Repeating a keyword, a sentence, or a paragraph emphasizes an idea in a powerful way. In engineering writing, it is better to consistently repeat the elements than to use synonyms. On the other hand, aimless repetition may impact conciseness because the sentence or paragraph might appear awkward and pointless. Therefore, you need to be prudent with repetition technique.

Example 12.25 shows that changing words from *reports* to *studies* to *analyses* might help with variety, but not for clarity or emphasis. These three words are slightly different in meaning: analyses are part of the studies that lead to the reports.

Example 12.25 Emphasis by Element Repetition

Poor repetition:
Recent reports concur our conclusions in this paper. These studies confirm that it is feasible to reduce greenhouse gas emissions by our technology. Their analyses, however, are primarily computational models.

Effective repetition:
Recent studies concur our conclusions in this paper. These studies confirm that it is feasible to reduce greenhouse gas emissions by our technology. Theirs, however, are primarily computational models.

12.8.4 Redundancy Due to Emphasis

Sometimes emphasis may cause redundancy, although it may enhance clarity. For instance, emphasis using intensifiers is unnecessary in engineering academic writing. *See* Sect. 12.10.

Emphasis by repeating the same idea, using the same or different words, also causes wordiness. You can emphasize by repetition in informal verbal communication, but you should avoid this practice in formal academic writing. For example:

Example 12.26 Redundant Emphasis

Informal/redundant   <u>At its core</u>, climate change is caused <u>primarily</u> through pollution, specifically the greenhouse effect.

Formal/concise   Climate change is caused <u>primarily</u> through pollution, specifically the greenhouse effect.

12.8.5 Garbled Sentences

Garbled sentences lose emphasis because all ideas are presented with equal importance. Loading a sentence with multiple thoughts reduces both clarity and conciseness of writing. Example 12.27 shows how to rewrite a garbled sentence.

Example 12.27 Garbled Sentence

Garbled   The company was founded in 2019, only three members were on the staff, and all members took multiple positions and roles, but it was not an efficient operation.

Revised   When the company was founded in 2019, only three members were on the staff, and all members took multiple positions and roles. However, it was not an efficient operation.

12.9   Modifiers

A modifier is a describer. It can be a word, a phrase, or a clause. You do not need modifiers to construct grammatically correct sentences, but you often need modifiers to provide details and to enhance clarity. It is especially useful to technical writing because many modifiers (words, phrases, or clauses) transform general descriptions into specific ones. Most modifiers *function* as adjectives that impose constraints on the words they modify, for instance, *ten* replications and *nanofiber* size.

Example 12.28 Modifier in Sentence

Unmodified   Sizes increased with potential.

Modified   <u>Nanofiber</u> sizes increased with the <u>electrical</u> potential <u>at the needle tip</u>.

12.9.1 Prepositional Phrases as Modifiers

A prepositional phrase can act as a modifier. It is normally composed of the preposition, its object, and the object to modify. However, you need to avoid awkward prepositions in sentence construction.

> **Example 12.29 Prepositional Phrases as Modifiers**
>
> - You should write an article <u>following the ethics guideline</u>. *[The preposition "following the ethics guideline" modifies the verb "write."]*
> - The results ~~are where they~~ were presented above.

12.9.2 Restrictive and Nonrestrictive Modifiers

Phrase and clause modifiers may be *restrictive* or *nonrestrictive*. A *restrictive* modifier restricts the meaning of the element it modifies. Otherwise, the modifier becomes *nonrestrictive*. Omission of nonrestrictive modifiers does not affect the main idea of the sentence.

> **Example 12.30 Nonrestrictive Modifier**
>
> The coronavirus disease 2019 (COVID-19<u>), which</u> was officially named as severe acute respiratory syndrome coronavirus (SARS-CoV-2), quickly spread around the world in early 2020.
>
> **Comment**:
> This sentence means the same as "The coronavirus disease 2019 quickly spread around the world in early 2020." The nonrestrictive modifier sets off by commas provides nonessential but extra information.

The same sentence may have different meanings, depending on whether the modifier is restrictive or nonrestrictive. Pay close attention to the commas that set off the modifier to avoid misleading sentences.

> **Note: which/that**
>
> In American English, *that* is used to introduce restrictive clauses, and *which* introduces nonrestrictive clauses. In British English, however, *which* can introduce both restrictive and nonrestrictive clauses. A non-native writer in English can always use *which* with nonrestrictive clauses but use *that* with restrictive clauses. It helps avoid confusion to the readers who are only familiar with American English.

12.9.3 Placement of Modifiers

A modifier should be placed immediately before or after the sentence element to be modified. This may be challenging to non-native speakers of English. Example 12.31 shows how placement of modifiers changes the ideas of sentences.

Example 12.31 Placement of Modifiers in Sentences

Words as modifiers:

- He almost lost all the data collected last night. [*Almost* modifies *lost*; *the data were not lost.*]
- He lost almost all the data collected last night. [*Almost* modifies *all*; *data* were lost but some survived.]

Clauses as modifiers:

- We sent the samples to the lab that we are satisfied with for analysis.
- We sent the samples that we are satisfied with to the lab for analysis.

Misplaced modifiers cause ambiguity. A squinting modifier is one type of misplaced modifier. A squinting modifier is out of place because it could modify two elements simultaneously. Nonetheless, you can correct squinting modifiers simply by rewriting the sentences.

Example 12.32 Squinting Modifiers

Squinting   They decide immediately to start the investigation. *[The word "immediately" could modify "decide" or "start."]*
Clear   They immediately decide to start the investigation.
Clear   They decide to immediately start the investigation.

12.9.4 Dangling Modifiers

A dangling modifier has nothing to modify; it typically appears at the beginning or the end of a sentence. Dangling modifiers occur when writers get ahead of themselves; they assume that the context is obvious to the readers and forget to help the readers get ready with the context.

Like squinting modifiers, dangling modifiers confuse the readers, but for a different reason. As seen in Example 12.33, you can correct dangling modifiers by adding appropriate subjects to the modifiers or the clauses.

Example 12.33 Dangling Modifiers

Dangling   <u>While working on the lab report</u>, the computer suddenly shut down *[It is not clear who is working. A computer cannot work on a report and the writer assumed the reader to know that it is "I".]*

Clear   <u>While I was working on the lab report</u>, the computer suddenly shut down.

Dangling   The figure becomes more readable <u>after replacing the symbols with descriptive words</u>. *[Who replaced the symbols?]*

Clear   The figure becomes more readable <u>when you replace the symbols with descriptive words</u>.

Clear   The figure becomes more readable <u>when the symbols were replaced with descriptive words</u>.

Dangling   Instead of filtration, Chu et al. added optically induced DEP after negative IMS. *[The phrase "Instead of filtration" is ambiguous: is it for DEP or IMS?]*

Clear   Chu et al. added optically induced DEP, instead of filtration, after negative IMS.

Clear   Chu et al. added optically induced DEP after negative IMS, instead of filtration.

12.10   Intensifiers and Subjective Writing

Intensifiers are the *adverbs* that emphasize degree. Unlike a modifier, which modifies works or phrases, an intensifier amplifies the meaning of the word (adverb or adjective) that it modifies. For example, *absolutely, best, extremely, much, rather, really, too,* and *very* are intensifiers.

Intensifiers function as comparison, especially in verbal communication. For example, "It was a *really* good party!" However, the use of intensifiers in engineering academic writing implies emotion and causes redundancy.

Overuse of subjective intensifiers implies emotion and undermines the professionalism of writing. For example, the adverbs *surprisingly*, *sadly*, *obviously*, and *definitely* in Example 12.34 intensify the clauses that follow them. The incorrect sentences impose personal feelings or opinions on the readers without concrete evidence. However, the readers may not feel the same way as the writer does.

Example 12.34 Avoid Subjective Writing

Incorrect   <u>Surprisingly</u>, they did not report the error analysis in their publication.

Correct   ~~Surprisingly,~~They did not report the error analysis in their publication.

Incorrect   We are <u>definitely</u> certain that it was the catalyst fouling which caused the damage.

Correct   We are ~~definitely~~ certain that it was the catalyst fouling that caused the damage.

Incorrect   <u>Obviously</u>, clean air is <u>absolutely</u> necessary for a healthy and productive economy.

Correct   ~~Obviously,~~ Clean air is ~~absolutely~~ necessary for a healthy and productive economy.

Therefore, you should avoid overuse or misuse of intensifiers in engineering academic writing. Overusing qualitative intensifiers (*very fine, too high, much longer*) may also cause vagueness and inaccuracy. Other intensifiers like *absolutely, perfectly, impossible*, and *final* cannot be used in academic publication because scientific research is continuously advancing. For example, the word *absolutely* is a misused intensifier in engineering writing.

A sentence often becomes concise and clear after we remove the unneeded or misused intensifiers. Chapter 7, for example, explains the importance of supporting arguments with numerical data instead of qualitative subjective descriptions.

Example 12.35 Misuse of Intensifiers

Incorrect   Clean air is absolutely necessary to our health.
Correct   Clean air is ~~absolutely~~ necessary to our health.
Incorrect   The test results and the model agree <u>very well</u>.
Correct   The error between the experimental and model results is <u>5% or less</u>.

12.11 Rambling Sentences

Section 10.3 emphasizes writing with smooth pacing. On the contrary, rambling sentences affect the pace of writing by overloading the readers with more information than they can comfortably take. Example 12.36 shows how to divide a rambling sentence into multiple simple sentences, starting with the main idea.

Example 12.36 Rambling Sentence

Rambling sentences:
In a CSTR, four dispersion patterns (flooding, cavity formation, complete dispersion of gas, recirculation of gas- liquid mixture) can be observed and the minimum impeller speed for complete dispersion of gas phase, N_{cd} is a key parameter considering the fact that gas-liquid contacting at impeller speeds below N_{cd} leads to a poor reactor performance as the lower part of reactor is wasted due to the incomplete gas dispersion. (Yu H. 2013, PhD Thesis).

Revisions to reduce rambling:
There are four dispersion patterns, flooding, cavity formation, complete dispersion of gas, and recirculation of gas-liquid mixture, in a CSTR. A minimum impeller speed is needed for a complete dispersion of gas phase. A lower speed leads to incomplete gas dispersion at the lower part of the reactor.

12.12 Practice Problems

Question 1   Choose the right verb.

1. The same data <u>is/are</u> plotted on in the same graph.
2. Either Table 1 or Figures 4.6 <u>supports/support</u> the argument.

Question 2 Rewrite these sentences using verb-based style.

3. Multiple hybrid strategies can be applied to improve biochemical methods.
4. As a result, supplemental DEP with pre-processing can be helpful since the cell density has a great impact on the enrichment performance.
5. DNA aptamers were linked to BSA for capturing CTC.

Question 3 Revise these sentences for better parallel structure.

6. Nanostructure adhesion of CTCs without labeling is a new and promising method owing to its simple operation and convenience for direct observation and cell culture.
7. Later, Ma et al. [73] fabricated three kinds of polystyrene fibrous substrates, i.e., microfibers (diameter of 4 6 μm), nanofibers (500–900 nm), and composites of nanofibers (100–650 nm) and microbeads (1–12 μm) for CTC isolation.
8. Under the influence of magnetic field, the labeled CTCs experienced lateral migration while unaffected blood cells were swept by buffer flow.
9. The IMS can be performed either in tubes (or wells) or by microchannels.
10. This research will integrate several academic disciplines, including the estimation of mobile source emissions of over thousands of air toxics, the atmospheric dispersion and deposition of these toxics, the fate and transport of toxics through various media, the assessment of the dose on humans, effects of toxic exposure on ecological receptors, the cumulative toxicological risks based on the MSATs doses, including in the analysis the uncertainty and stochastic nature of these processes. *[Curtsey: M. Munshed].*
11. These types of infectious microorganisms include pneumonic plague, Legionella pneumophila, tuberculosis, influenza viruses, herpes viruses, SARS virus (a form of coronavirus), pathogenic streptococci and staphylococci from saliva and nasopharyngeal secretions, HIV, and hepatitis B and C viruses.
12. LIB has been widely used from portable electronics, electrical vehicles to large-scale power sources. (Li and Tan 2020).
13. The continuous researches and developments in all aspects of LIB materials, including electrodes, electrolyte, separator, and current collectors, keeps improving the battery cost efficiency, energy, safety, and power capability.
14. According to the dependent property of CTCs or hematologic cells, CTC enrichment techniques with single modality can be classified into biophysical methods and biochemical methods.

Question 4 **(Conjunctions/transitions)** Highlight the conjunction words in the sentence. Understand their functions in ideas development by reading the paragraphs with and without transition.

15. The analysis of the organic fraction of atmospheric aerosol particles on a molecular level is often challenging because their composition can be highly variable in time and space and usually only small sample amounts are available for analysis because atmospheric concentrations are typically a few micrograms per cubic meter. The composition also depends on the size fraction of the aerosol; for example, resuspended, wind-blown particles are mostly larger than 1 μm and have often a very distinct chemical composition compared to particles smaller than 1 μm, which are mainly emitted by combustion sources or formed by chemical reactions in the atmosphere. Thus, analytical−chemical techniques used to analyze atmospheric organic aerosol particles need to be highly sensitive to allow for highly time-resolved analyses at trace concentrations. In addition, the analysis technique needs to be able to characterize highly complex organic compound mixtures with thousands of mostly unknown components (Fuller et al. 2012).

16. There are a number of analytical—chemical techniques that allow measuring the composition of aerosol particles with high time resolution. Online aerosol mass spectrometry (AMS) techniques, for example, allow for very highly time-resolved particle composition studies. However, such measurements are demanding with respect to manpower and other resources and are usually performed only for a few weeks at a specific site. Thus, such measurements rarely provide insight into long-term trends of particle composition. In contrast, aerosol samples collected on filters or impactors over long time periods are more readily available, but have mostly a rather low time resolution of the order of a day or more, and their chemical analysis usually involves time-consuming workup and is prone to artifacts during sample preparation (Fuller et al. 2012).

17. Nowadays, the most popular online dating Web applications could even have several hundreds of millions of registered users. Consequently, an effective reciprocal recommendation system is urgently needed to enhance user experience. Generally, the reciprocal recommendation problem aims to recommend a list of users to another user that best matches their mutual interests. For example, in an online dating platform (e.g., Zhenai 1 or Match 2), the purpose of reciprocal recommendation is to recommend male users and female users who are mutually interested in each other. *[Source:* https://arxiv.org/pdf/2011.12586.pdf*]*.

18. A user might send a message to another user if and only if the other user has certain content of profile that is preferred by the user, denoted as user's preferred attribute. On the contrary, if a user does not reply to a message, it indicates that either there are no preferred attributes or there is at least one attribute of the other user that the user does not like, which is called repulsive attribute in this paper. For example, user A with a good salary may prefer user B (to be recommended) having a decent occupation; whereas user P who has children may dislike the drinking or smoking user Q. Thus, occupation is a preferred attribute of user B to user A, and drinking or smoking is a repulsive attribute of user Q to user P. Moreover, the salary - occupation forms a preference interaction between a pair of users, while children - drinking and children - smoking form the repulsiveness interaction. Obviously, different users may have different sets of preferred or repulsive attributes. Hereinafter, we call these attributes the key attributes to avoid ambiguity. [*Source:* https://arxiv.org/pdf/2011.12586.pdf].

19. The objective of this study is to investigate the size-resolved filtration efficiency of nanofibrous media in an actual air cleaner. To this end, three electrospun nanofibrous media mounted in mechanical portable air cleaners and challenged with nanoparticles generated by an ultrasonic essential oil diffuser. The filtration efficiencies of these nanofibrous media were also determined in a more conventional duct test when filter medias were installed in a cone-shape filter holder and were challenged with a salt aerosol. We investigated the effects of different experimental setup, sample variation, filter media, face velocity, electrical charges, and dust loading on filtration efficiency. The overall goal of this study is to draw links between air cleaner and laboratory-scale results. (Givehchi et al. 2021).

20. The mass of dust collected on the media for this duration was 25.7 ± 3.7 mg. These dust particles may change the morphology and diminish the favorable features of the nanofibers. The accumulated mass also decrease the higher pressure drop and lower the face velocity. Earlier studies also show that particle loading in nanofibrous media increases the pressure drop more rapidly than micron-fibers. Meanwhile, the face velocity decreased from 32 cm/s to 24 cm/s due to particle loading. The lower the face velocity is, the higher the filtration efficiency for smaller particles due to increased residence time for diffusion, and the lower the filtration efficiency for larger particles due to reduced inertial impaction. Therefore, our finding suggests that dust loading adversely compensates the effects of lower face velocity for smaller nanoparticles, which lowers the overall filtration efficiency.

Question 5 (Emphasis) Change the order of elements in the following sentences by beginning with the important ideas. (The first sentence of a paragraph states the controlling idea of the paragraph.)

21. After incubation of blood sample with antibody-coated magnetic beads in a syringe to label CTCs, magnetic separation followed by rinsing was carried out three times. Then the treated sample was filtered by a membrane with pore size of 5 μm. In tests of enriching MCF-7 (10–300 cells) from 0.5 mL blood, recovery efficiencies 70–90% were obtained.

22. When doped with 270 nm PDA nanospheres, the "necklace" nanostructure yielded best capture efficiency, 50–70%, for 4 different kinds of cancer cells.

23. Because their composition can be highly variable in time and space, the analysis of the organic fraction of atmospheric aerosol particles on a molecular level is often challenging. (Fuller et al. 2012).

24. Because atmospheric concentrations are typically a few micrograms per cubic meter, usually only small sample amounts are available for analysis. (Fuller et al. 2012).

25. Because of the tiny diameters involved, the *Reynolds numbers* for such flows are extremely small and the flow is definitely laminar. (Munson et al. 2009).

26. Due to breakdown of the liquid junction, contact times of over 30 s are less effective.

27. To aid mixing of the extracted sample into the solvent within 5 s, many previous studies using LESA have repeatedly deposited and aspirated solvent onto the sample on a single extraction spot. (Fuller et al. 2012).

28. In order to establish a quantitative relationship between winter road condition and vehicular air emissions, multiple linear regression models were developed. (Min 2015).

29. When U is taken as the control parameter, the BDs for Δ = 0.0, 0.001, 0.005 are shown in Fig. 8. [*Source:* www.quman.org/article/article0167.html].

30. Based on the *triangulation structure* built from unorganized points or a CAD model, the extended STL format is described in this section. [*Source:* www.quman.org/article/article0167.html].

31. To evaluate the models, we adopt five popular evaluation metrics, i.e., Precision, Recall, F1, ROC, and AUC and the threshold is set to 0.5 for precision, Recall and F1. [*Source:* https://arxiv.org/pdf/2011.12586.pdf].

32. When a power plant plume is sampled far from the source, it has been diluted thousands of times with air from the surroundings. In order to determine the contribution of the source to levels in the plume, the mass contributed by the background air must be subtracted. Because of the high levels of dilution, it is critical that the background be accurately characterized. (Imhoff et al. 2000).

33. To reduce nonspecific adhesions of blood cells on nanofiber substrate, zwitterionic pCBMA with antifouling effects was employed.

34. However, there are also several drawbacks for the multi-step methods. Most multi-step methods can realize high capture efficiency (>90%) by the superimposed and enhanced capture effects, but the loss of CTCs in waste outlets by multiple enrichment units lowers down the efficiency to ~80%.

Question 6 (Modifiers/intensifiers) Revise these sentences by reducing subjective, emotional tones.

35. Developing an automated CTC enrichment platform showing high capture efficiency, purity, and cell viability is undoubtedly preferred.

36. The TiO_2 nanofiber platform successfully detected CTCs from 7 out of 7 blood samples only containing 3–19 CTCs.

37. Unfortunately, due to the generally low peak intensities of the identified species, MS/MS analysis for further structural identification was not possible. (Fuller et al., 2012).

38. It is, of course, impossible to summarize the rich history of fluid mechanics in a few paragraphs. Only a brief glimpse is provided, and we hope it will stir your interest. (Munson et al. 2009).

39. Surprisingly, there are two real, positive roots.

40. Obviously, which characteristic area is used in the definition of the lift and drag coefficients must be clearly stated. (Munson et al. 2009).

41. The results stated above reveal that the choice of drag models can significantly affect the simulated flow patterns. It seems that both under- and over-estimations of drag coefficients in the dense phase bring about the deviation of the flow pattern from stable spouting operation, resulting in a bubble/slug flow pattern. For dense phase systems, van Wachem et al. (2001) showed that the Syamlal and O'Brien (1988) model led to lower predictions of pressure drop and bed expansion, in good agreement with our results. (Du et al. 2006).

42. The fast drop indicated more significant effect of the separator thickness on the battery energy density than the separator porosity. (Li and Tan 2020).

43. Since gravity is the driving force in these problems, Froude number similarity is definitely required. (Munson et al. 2009).

44. Because of the tiny diameters involved, the Reynolds numbers for such flows are extremely small and the flow is definitely laminar. (Munson et al. 2009).

45. The centrifugation is a fairly mature method for CTC enrichment.

Question 7 Rewrite these sentences to improve clarity.

46. Single-modality enrichment methods depending on either biophysical or biochemical property difference between CTC and hematologic cells face with limitations and haven't been broadly adopted for clinical utility.

47. Moreover, this is still a relatively new method with the first systematic study reported in 2013 and research in this area still focuses on characterization of nanoroughened surfaces (e.g., stiffness, morphology, etc.) and interactions between tumor cells and nanostructure (e.g., guiding effect, focal adhesion, etc.).

Phrases and Words

13

13.1 Phrases

A phrase is a group of words. A meaningful phrase acts as an adjective, adverb, noun, or verb in a sentence. Unlike clauses, phrases cannot state an idea because it contains neither a subject nor a predicate.

Table 13.1 shows examples of nouns, adjectives, verbals, and verbs with the same roots. They form adjective phrases, adverb phrases, noun phrases, gerund phrases, verb phrases, and so on. Example 13.1 illustrates adjective and adverb phrases, and the rest of this section introduces other phrases.

Table 13.1 Words with the same roots

Noun	Adjective	Adverb	Verb
consideration	considerate	considerately	consider
description	descriptive	descriptively	describe
explanation	explanatory	explanatorily	explain
production	productive	productively	produce
detail	detailed	–	detail

Example 13.1. Adjective Phrase and Adverb Phrase

Adjective Phrase

- This item has been received already. (*without phrase*)
- This item <u>on the list</u> has been received already. (*with phrase*)

Adverb Phrase

- We have received this item on the list <u>in perfect shape</u>.

© The Author(s), under exclusive license to Springer Nature Switzerland AG 2022
Z. Tan, *Academic Writing for Engineering Publications*,
https://doi.org/10.1007/978-3-030-99364-1_13

13.1.1 Noun Phrases

A noun phrase functions as a subject, object, or prepositional object in a sentence. It is composed of a noun and its modifiers.

> **Example 13.2. Noun Phrases**
>
> Before the <u>publication of your articles</u>, you may have to sign the agreement to release your copyrights.

13.1.2 Gerund Phrases

A gerund phrase is originated from a verb ending with *-ing*. It is different from a noun phrase. A gerund phrase can act as either a subject or an object in a sentence (Alred et al. 2018).

> **Example 13.3. Gerund Phrases**
>
> **Subject:** <u>Writing this book</u> is driven by my enthusiasm in higher education.
> <u>Writing a journal article</u> requires certain skills and good planning.
> **Object:** All engineering researchers enjoy <u>writing journal articles.</u>

13.1.3 Verb Phrases

A verb phrase is formed by an *auxiliary verb* followed by the *main verb*. Some words may appear between the auxiliary verb and the main verb. In the following sentence, for example, *am* is the auxiliary verb, and *writing* is the main verb.

> **Example 13.4. Verb Phrases**
>
> * I <u>was writing</u> this book when the school was closed because of COVID-19 pandemic.
> * This item on the list <u>should have been received</u>. (*In reality it has not.*)

13.1.4 Prepositional Phrases

Prepositional phrases normally modify nouns or verbs in the same sentence. They may act as adjectives following the nouns that they modify. *See* Example 13.5.

> **Example 13.5. Prepositional Phrases**
>
> * The power will turn off automatically <u>after the temperature reaches</u> 350 °C.
> * Agricultural biomass waste <u>with a high cellulose content</u> can be converted into biofuel by hydrothermal conversion.

On the other hand, Example 13.6 highlights that overuse of prepositional phrases can lead to wordiness or ambiguity. For clarity, you can use modifiers, which are more economical, for the same purpose as prepositional phrases.

Example 13.6. Prepositional Phrases and Modifiers

Wordy: The data presented <u>with solid black circles on the dashed lines in Figure 3</u> are obtained experimentally <u>in the lab</u>. (*prepositional phrases used*)

Concise: The data in Fig. 3, presented using <u>solid black circles on the dashed line</u>, are experimental results. (*modifier*)

Ambiguous: The curve for model and that for experiment <u>in red color</u> agree with each other. (*whether one or both in red*)

Clear: The curve for model <u>in red color</u> and that for experiment agree with each other.

13.1.5 Participial Phrases

A participial phrase is a verbal that ends in *-ing* or *-ed*. It functions as an adjective in sentence. For clarity, it is important to describe the relationship between a participial phrase and other elements in the same sentence. Otherwise, the phrase becomes dangling and modifies nothing in the sentence. Many new writers overuse participial phrases in academic writing. As a result, the sentences become fragmented or dangling. It may also cause ambiguity and vagueness. *See* Example 13.7.

Example 13.7. Participial Phrases

Dangling: The system <u>having the highest thermal efficiency</u> chosen for further studies at the pilot scale.

Correct: The system <u>having the highest thermal efficiency</u> was chosen for further studies at the pilot scale.

Dangling: <u>Being unsatisfied</u> with the accuracy, <u>the device</u> was replaced with another one. (*Dangling phrase: The device cannot be unsatisfied; it must be person.*)

Correct: <u>Being unsatisfied</u> with the accuracy of the data, <u>we</u> replaced the device with another one for further studies at the pilot scale.

13.1.6 Infinitive Phrases

An infinitive phrase begins with the word *to*, followed by a verb. It indicates the purpose of an action. In comparison, a prepositional phrase also begins with *to* but followed by a noun or a pronoun. A prepositional phrase normally indicates a destination for an action.

Example 13.8. Infinitive and Prepositional Phrases

Infinitive: <u>To meet the deadline</u>, I am willing <u>to work overtime</u> every day for the coming week.

Prepositional: I go <u>to school</u> for education.

Both: I go <u>to the library to prepare</u> for the final exam.

13.2   Verb-Based Writing

Use verb-based expression when you can, although the preceding section shows that various phrases are available in English. Avoiding other styles, especially noun-based styles, greatly enhances the conciseness of engineering academic writing. Sometimes, verbal-based styles is also acceptable.

To achieve verb-based writing, you can convert noun-, adjective-, and adverb-based styles to verb-based style because they often have the same root (*see* Table 13.1). Example 13.9 compares different styles of writing. When you revise your draft manuscript, you can underline each verb or verb-derived words and revise them following these guides:

- Remove expletive openers such as *there is*, *there are*, and *it is*, if any, and replace *to do* with a strong verb.
- Replace nouns ending in *-tion*, *−ment*, and the like with their main verbs.
- Change passive to active voices using doer.
- Convert verbal phrases into their strong verbs.
- Turn piles of prepositions into possessives, relative clauses, or adjectives.

Example 13.9. Using Verb-Based Writing Style

Expletive: There are many researchers who have reported the effects of air pollution on the environment and public health.

Noun-based: Many researchers have reported the effects of air pollution on the environment and public health.

Expletive: It is reported that the deposition of non-noble metal Bi12–14 on the surface of semiconductors improves photocatalytic activity.

Noun-based: The deposition of non-noble metal Bi12-14 on the surface of semiconductors improves photocatalytic activity.

Verb-based: Depositing non-noble metal Bi12-14 on the surface of semiconductors improves photocatalytic activity.

Expletive and noun-based: There is a need of cost-effective technologies for nitric oxide emission control.

Verb-based: The power industry needs cost-effective technologies for nitric oxide emission control.

Adjective-based: A machine is more productive than a worker.

Verb-noun based: A machine enhances a worker's productivity.

Verb-based: A machine produces faster than a worker.

13.2.1 Phrasal Verb Error

A phrasal verb is composed of a verb followed by an adverb or preposition or both. You should avoid phrasal verbs in formal writing because of their informality. A phrasal verb may also confuse non-native English readers because the meaning of the phrasal verb is different from either the verb or the adverb or preposition that follow. Nonetheless, you often find a formal equivalent of a phrasal verb, and the formal alternative is usually short or single-worded. (*See* Example 13.10.)

Table 13.2 Comparison of phrasal verbs and verbs

Phrasal verb	Verb
agree to	consent
call off	cancel
crop up	arise/emerge
do over	repeat
drop off	deposit/deliver/left
find out	discover
follow through	persist
have a great impact on	greatly impact
get it	understand
get rid of	remove
get through with	finish
hold on	pause/wait
is in agreement with	agrees with
keep on/keep up	continue
keep track of	calculate/record
leave out	omit
let off/let go	release/lightly punish
look up	Study
make *A* stronger	strengthen/reinforce *A*
put forward	propose/suggest/nominate
rule out	eliminate/reject
run out of	deplete
take into consideration	consider
tone down	soften/moderate

Example 13.10. Phrasal Verb vs. Concise Verb

Phrasal verb: I <u>dropped off</u> the samples at the lab this morning.
Neutral verb: I <u>left</u> the samples in the lab this morning.
Formal verb: I <u>delivered</u> the samples to the lab this morning.

Table 13.2 compares some phrasal verbs vs. regular verbs that are often used by mistake in engineering publications. You can compile yours for your future reference. Then you can replace the phrasal verbs with simple verbs in your writing.

13.2.2 Nominalization Error

Nominalization is another mistake that non-native English writers frequently make in verb-based writing. As shown in Table 13.3, a nominalization is a noun combined with a vague verb like *cause*, *do*, *give*, *make*, or *perform*. For example, *give an introduction* is a nominalization of *introduce*.

Nominalizations cause awkwardness in reading, although the idea remains clear. Example 13.11 compares nominalizations with their strong verbs for the same ideas. Strong verbs enhance the conciseness of writing and improve reading experiences.

13.3 Wordiness

Your writing can become wordy when you use more words than necessary. As a result, the writing ends up lacking effectiveness and conciseness. There are many reasons for wordiness, including phrasal verb and nominalization as explained in the preceding section. Others include vague modifiers, unneeded intensifiers, redundancy, duplication, and stocking phrases. The rest of this section explains these wording errors in engineering writing.

13.3.1 Duplicated Nouns

Example 13.11. Nominalization vs. Strong Verb

Wordy: Chapter 2 <u>gives an introduction of</u> the methods used in this study.
Concise: Chapter 2 <u>introduces</u> the methods used in this study.
Wordy: We <u>performed a study on</u> the effects of air pollution on respiratory diseases.
Concise: We <u>studied</u> the effects of air pollution on respiratory diseases.
Wordy: <u>Avoid using</u> subjective words in formal writing.
Concise: Avoid ~~using~~ subjective words in formal writing.
Clear: Avoid <u>using</u> subjective words <u>as modifiers</u> in formal writing.

Table 13.4 lists some duplicated nouns collected from references in this book and students' drafts. You can use either the correct one or the acceptable one, but not both.

Table 13.3 Nominalization

Nominalization	Concise verb
cause confusion to	confuse
conduct research	study
give an introduction	introduce
make revisions	revise
make an adjustment of	adjust
perform a study	study
perform numerical simulation	simulate, model (*as verb*)
provide details	elaborate
take into consideration	consider

Table 13.4 Duplicated nouns

Duplicated nouns	Correct	Acceptable
application results	results	application
arrangement plan	arrangement	plan
calculation results	results	calculation
knowledge memory	knowledge	memory
main focus	focus	–
output performance	output	performance
research work	research	work
simulation results	results	simulation
sketch drawing	sketch	drawing
schematic diagram	schematic	diagram
red color	red	–

13.3.2 Redundancy

Table 13.5 lists some redundant phrases, which often appear in draft manuscripts that I revised. Redundancy occurs when multiple words with the same meaning appear in the same phrase. Repeating the same idea for emphasis sounds normal in verbal communication, but it may annoy readers by reducing conciseness of formal writing. Example 13.12 shows that you can replace redundant phrases with single words and rewriting the sentence.

13.3.3 Avoiding Wordy Phrases

Table 13.6 is an incomplete list of wordy phrases that my new graduate students use in their rough drafts. Many non-native English speakers pay more attention on grammar and spelling than conciseness. As a result, they use unnecessarily long but grammatically correct phrases. You can often replace these long phrases with short ones or single words.

Example 13.12. Redundancy in Writing

Redundant: The probe's <u>length</u> is 20 cm <u>long</u>.
Concise: The probe's length is 20 cm.
Concise: The probe is 20 cm long.

Table 13.5 Redundant wording

Redundant	Concise
a flame that is red in color	a red flame
actual experience	experience
all of	all
a brand new innovation	an innovation
basic and fundamental	fundamental
basic essentials	essentials
by using	by/with/using
collaborate together	collaborate
completely accurate	accurate
completely finished	finished
cylindrical in shape	cylindrical
each and every	each, every
eight hours of time	eight hours
first and foremost	foremost
final outcome	outcome
inside of	inside
later on	later
other alternative	alternative
perfectly clear	clear
present status	status
two cubic meters in volume	two cubic meters
uniformly homogeneous	uniform
visibly see	see

Table 13.6 Long phrases vs. concise words

Stocking phrase	Concise word
a large number/quantity of	numerous/many
a reactor designed and constructed in-house by the authors	a home-made reactor
A is located in the immediate vicinity of B	A is near B
As seen in Figure 2,…	Figure 2 shows that…
at a rapid rate	rapidly
at this point in time	now, at this point
at this precise moment in time	now
at this time	now, currently
be likely to do	likely do
by means of	by
do not hesitate to	please
due to the fact that	because
during the course of	during
exhibits the ability to	can
for the reason that	because
He is of the opinion that	He thinks
I am in receipt of	I have
in a position to	can
in consequence	so
in excess of	more than
in many cases	often
in respect of	about
in the event that	if
in the first place	first
in the form of	as
in the event of	if
in the situation of doing	can do
in the near future	soon
is/are found to be	is/are
is located next to	s next to
It can be seen that	(*Remove this phrase*)
It is likely that A will arise with regards to the completion of B	You will probably have A completing B
of a satisfactory nature	be satisfactory
of a satisfactory character	be satisfactory
owing to the situation that	because
owing to the fact that	because
prior to the occasion when	before
should a situation arise where	if
taking into consideration such factors as	considering
There is a reasonable expectation that	Probably
through the usage of	by, with
two or more	multiple
with regard to	regarding/about
with the exception of	except
within a comparatively short period	soon
Of particular interest,	Particularly,
the best operating condition for 5-HMF production with the maximum yield	the optimal condition for 5-HMF production
In order to be able to use solid powder catalysts…	To use powderous catalysts
study the performance of a series of porous niobium phosphate solid acid catalysts	study a series of porous niobium phosphate solid acid catalysts
the acidic property of	the acidity of

Particularly, Sect. 13.2 introduces the wordiness due to unneeded sentence openers like *There is*, *It is…to*, *It is can be seen that*, and *As seen in* (*see* Example 13.9). In addition, converting independent clauses introduced by *which is*, *who are*, and the like into a short expression also helps improve conciseness. *See* Example 13.13.

Example 13.13. Wordiness Due to Independent Clause

Wordy: Thousands of professionals will be attending the annual <u>conference, which is scheduled</u> on the 25th of next June.

Concise: Thousands of professionals will be attending the annual <u>conference scheduled</u> on the 25th of next June.

13.4   Shift of Function

Shift of function occurs when the function of a word shifts from one to another, depending on the context. Shift of function occurs frequently easily, and it may be a challenge to non-native English speakers.

Example 13.14. Word Challenges (Functional Shift)

- <u>Talk</u> the <u>talk</u>, <u>walk</u> the <u>walk</u>. (*The first talk and the first walk are verbs; the second, nouns.*)
- <u>After</u> we present the results, we carry out in-depth analyses. <u>After</u> the analyses, we look <u>after</u> the conclusions. (*The word after shifts its function from conjunction to preposition.*)

13.5   Jargon, Slang, and Idiom

Jargons, slangs, and idioms are words and phrases used only locally. Such words are usually considered informal, and readers out of the regions are unfamiliar with their meanings because of the loss of context. For example, *sub* may mean *a long sandwich* in the USA (United States of America). It may also mean *submarine* or *subscription*, depending on the context. As a verb, it is shortened from *substitute*.

Example 13.15. Local Words

- "Want a <u>sub</u>?" "Yes, please!"
- Tom <u>subbed</u> for Scott as goalkeeper in this soccer game.

Jargon is specialized words or expressions that are used within a profession or group. Only the readers who are unique to an occupation understand the jargon. Jargon can also refer to terminology used by people in a specialized field. For example, *debug code* is a jargon in computer science, meaning test for errors or determine the cause of errors. Jargons occasionally help with conciseness, but they often reduce clarity of writing for readers worldwide.

Slang and idiom refer to a type of informal language used only in certain situations. Slang and idioms often lead to functional shifts of words and confusion to readers.

A slang may be a familiar word, but it has a special meaning that may be different from its literal meaning. For instance, *chill* can mean *relax* or *calm down*; *nerd* means a socially awkward person.

> **Note**
>
> Use words and phrases only when you know their precise meanings.

Like a phrasal verb (*see* Sect. 13.2.1), an idiom is a phrase that has a different meaning from its constituent elements. An idiom may function as a verb, but it can be much more than that. Among a variety of phrases, idioms may be as challenging as slangs to non-native speakers of English. Like slang, idioms are phrases that cannot be *literally* interpreted. For example, *run a test* means *to arrange for someone or something to be tested* – nobody is running, literally.

Table 13.7 lists some idioms commonly used in engineering communication. These idioms can make writing more vigorous, creating a natural connection between the writers and the readers. However, idioms and slang are culture specific. Non-native English speakers may need them for effective verbal communication but should avoid them in formal writing.

You should avoid jargon, slang, and idioms in formal writing. They may cause not only confusion but also even frustration to non-native English speakers. A non-native English reader may stop reading and check dictionaries to interpret the precise meaning of jargon, slang, and idioms.

13.6 Contractions

Contractions should only be used in informal verbal communication. You need to spell out the words in academic writing. Table 13.8 compares some contracts with their words – pay attention to *cannot*, which is one word.

Table 13.7 Examples of idiom

Idiom	Meaning
One's cup of tea	something one likes
agree with	in accord
cross out	draw a line through
watch out for	be careful

Table 13.8 Contraction examples

Contraction	Full word
aren't	are not
can't	cannot
don't	do not
isn't	is not
haven't	have not
won't	will not

13.7   Vague and Subjective Words

You should avoid vague and subjective words in engineering academic writing. Vague words have different meanings and interpretations. Subjective words can also have vague meanings. Table 13.9 lists some vague and subjective words that you should avoid in academic writing. You use these words as modifiers or intensifiers only when it is necessary. In addition, they should be followed with specifics when you choose to use them.

Example 13.16 compares vague and subjective writing with clear and specific writing. The vague and subjective words *significant, best, critical,* and *subtle* are intensifiers in this paragraph. This abstract becomes subjective to the readers because it lacks supporting evidence.

Words like *a lot, good, best,* and *subtle* are often vague words. *See also* subjective writing, modifiers, intensifiers, etc.

Example 13.16. Vague and Subjective Words

The quantitative analyses showed that the different drag models led to <u>significant differences</u> in dense phase simulations. Among the different drag models discussed, the Gidaspow model gave the <u>best agreement</u> with experimental observation both qualitatively and quantitatively. The present investigation showed that drag models had <u>critical and subtle impacts</u> on the CFD predictions of dense gas–solid two-phase systems such as encountered in spouted beds. (*Source:* Du et al. 2006)

13.7.1 Using the Word *Significant*

A commonly misused word is *significant* (as an adjective) or *significantly* (as an intensifier). In engineering studies, the word *significant* or *significantly* is reserved for a conclusion supported with statistical analysis. *See* Example 13.17 (Fuller et al. 2012).

Example 13.17. Correct Use of *significant* with Statistics

"Linear regression analysis was used to indicate correlations, and any correlation with an R^2 value of greater than 0.8 was defined as <u>statistically significant.</u>"

Table 13.9 Vague and subjective words

Vague words	A lot, above, aforesaid, aforementioned, agree well, and/or, appreciable, approximate, bad, best, best agreement, comparative, definite, evident, excessive, fair, fairly, good, important, most, much, obviously, poor, reasonable, relative, significant, significantly, some time, sometime, sometimes, sufficient, suitable, tall, thing, rather, too high, utter, very, very fine, well…
Subjective words	absolutely, absurd, actually, as a matter of fact, bad, completely, definitely, entirely, extremely, fully, impossible, in fact, really, ridiculous, sadly, silly, surprisingly, perfectly, poor, too bad, unquestionably, unfortunately…

Without statistical analysis, however, *significant*, *significance*, or *significantly* conveys little information. Instead of imposing your personal opinion on your readers, you need to write down the values and allow your readers to decide its *significance*, for example:

Example 13.18. Misuse of *significant*

Vague: The accuracy of the model is improved <u>significantly</u>.
Clear: The accuracy of the model is improved <u>by 50%</u>.

13.7.2 Using the Word *Fact*

Many writers use the word *fact* in engineering academic writing, but phrases like *as a matter of fact* and *in fact* are subjective. The reason is that you cannot force your readers to accept your judgment or opinion as a fact. Even the observation and measurement may not be the facts because there are probable errors introduced by human factors or the devices. To revise, you can simply remove similar phrases. This removal also improves conciseness of your writing.

Example 13.19. Expressions Containing the Word *fact*

Incorrect: <u>As a matter of fact</u>, our research shows that solvent viscosity is a dominating factor for nanofiber fabrication by electrospinning.
Revised: Our research shows that solvent viscosity is a dominating factor for nanofiber fabrication by electrospinning.
Incorrect: Ligand affinity can be typically combined with biophysical method of filtration or centrifugation to develop multistep hybrid-modality methods for CTC enrichment, <u>due to the fact</u> that labeling CTC with particles by ligand affinity can improve the filtration and centrifugation efficiencies by increasing the size and density contrasts between CTCs and blood cells.
Revised: Multistep hybrid-modality methods combine ligand affinity with filtration or centrifugation; labeling CTC with particles by ligand affinity can improve the filtration and centrifugation efficiencies by increasing the size and density difference between CTCs and blood cells.

13.8 Pairs of Challenging Words

Pairs of words have similar meanings or functions, but not precisely the same. Such words exist in any language, and these words are known as *translators' false friends* in linguistics. Particularly, they can be challenging to non-native English speakers. This section comprises of pairs of challenging words. It is an incomplete list of words that are challenging to non-native English speakers including myself. You can build your own list by addition to or removal from the entries. (*Sources: Cambridge Dictionary; references in this book; students' rough drafts*)

a/an/the

The grammatical rules of articles are complex to many non-native English speakers. Typical mistakes include omission when they are required, wordiness when they are not needed, and incorrect locations where they do not belong. For example,

---

Example 13.20. Articles

Incorrect: A computation fluid dynamics (CFD) model for <u>air flow</u> around cylindrical object and investigation of <u>the air</u> velocity profile around cylinder is presented here. Turbulence is visualized to observe the effects of flows.

Correct: <u>Presented here is</u> a computation fluid dynamics (CFD) model for *<u>the</u>* <u>air flow</u> around *<u>a</u>* cylindrical object and *<u>a</u>* visualization of ~~the~~ air velocity around the cylinder. Turbulence is visualized to observe the effects of flows.

---

a lot/many/numerous

Do not use *a lot* in formal writing; instead, use *many* or *numerous* for estimation. It is even better to give specific values when they are available.

above/aforesaid/aforementioned

Above, *aforesaid*, and *aforementioned* are vague words typically used in legal documents. Avoid using these words to refer to the preceding nouns and pronouns. Simply repeat the nouns or pronouns for clarity (*see also* Avoiding Legal Words).

accuracy/precision

Accuracy means being correct without a single error. *Precision* is refinement in a measurement, calculation, or specification. For instance, 3.14159 is more precise than 3.14. *Precision* does not necessarily mean *accuracy*. *See* engineering statistics books for more information.

activate/actuate

Activate something mean *make something active*. *Actuate* is usually limited to mechanical devices.

adapt/adopt

Adapt means *to adjust to new conditions* or *to modify for a new purpose*. *Adopt* means *to accept* or *to take on*.

Example 13.21. Adapt/adopt

We must <u>adopt</u> the latest safety guideline to <u>adapt</u> to the new working environment.

---

affect/effect

Affect is a verb, and *effect* can function as both a noun and a verb. When *effect* acts as a noun, it means *result*. When *effect* is used as a verb, it means *result in*, *cause*, *make*, *produce*, and the like. Non-native English writers may want to avoid using *effect* as a verb, if possible, in technical writing.

all together/altogether

Altogether means *completely, everything considered,* or *on the whole. All together* means *all in one place, all in a group,* or *all at once.*

also/too/in addition

Also, too, and *in addition* have very close meanings. However, do not use *also* to open a sentence or to connect two elements in a sentence; instead, use *in addition* (which is a preposition). *Too* ends a sentence.

Example 13.22. Also/too/in addition

Incorrect: Heat exchangers are used in power plants, <u>also</u> they are used in other systems.
Incorrect: Heat exchangers arc used in power plants. <u>Also</u>, they are used in other systems.
Correct: Heat exchangers are used in power plants, <u>and</u> they are used in other systems.
Correct: Heat exchangers are used in power plants; they are used in many other systems too.
Correct: Heat exchangers are used in power plants. <u>In addition</u>, they are used in other systems.

---

amount /number/quantity

Amount is used for mass nouns that cannot be counted by number (cement, electricity, paper, water). *Number* is used with countable nouns (device, book). *Quantity* can be used with both mass nouns and countable nouns. *Quantity* is more formal than *amount* or *number*. For example:

Example 13.23. Amount/number/quantity

- The <u>amount of cement</u> is determined before it is mixed with sand.
- The <u>number of bags</u> of cement is determined before mixing the cement with sand.
- The <u>quantities of cement and sand</u> are determined before mixing them.

---

and/or vs. or...or...or both

Avoid *and/or* in formal writing; instead, use *or...or...or both*. For example:

Example 13.24. And/or

Confusing: You can use <u>WebEx and/or Zoom</u> for video conference.
Clear: You can use <u>WebEx *or* Zoom *or both*</u> for video conference.

---

as/because/since

As, *because*, and *since* are commonly used to introduce subordinate clauses that express causes. *Because*, more commonly used than *as* and *since*, introduces the most specific reason. *Since* is a weak substitute for *because* but stronger than *as*. However, *since* can be used for additional factors emphasizing on circumstance or time. Avoid using *as* to indicate cause in formal writing, although it can be used informally.

Example 13.25. As/because/since

Since the paper has been published, we must prepare a list of errata for the readers because there are several typos in the text. (*Since indicates change of circumstance; because, the cause of action.*)

---

as such/thus/therefore

Avoid the phrase *as such* in formal writing. Instead, use *thus* and *therefore*.

as well as/and

Either *as well as* or *and* means *in addition*, but they are used with different verbs in sentences.

Example 13.26. As well as/and

- Both the temperature and the pressure of an ideal gas can affect its volume.
- The temperature as well as the pressure of an ideal gas can affect its volume.

---

average/mean

Both *average* and *mean* are statistical terms; they are usually confused with one another, especially to non-native English writers. There are several types of *means* in statistics, and *average* (or arithmetic *mean*) is the simplest form of *mean*. Make sure you know precisely what kind of *means* you are referring to in your writing. *See* also *significant* as a modifier.

begin/start

Begin is an irregular verb, and *start* can act as a noun and a regular verb. *Begin* is more formal than start (as a verb), but they have the same meaning. *Start* and *beginning* mean the same thing when they are used as nouns.

beside/besides

Beside means *in a position immediately to one side of*; *besides* means *as well* or *in addition to*. Do not use *besides* as a conjunction. Instead, use *in addition* or *additionally*.

Example 13.27. Beside/besides

Besides the agitated heater, the oven and the water bath are located *beside* each other on the bench.

---

between/among/amongst

Between is normally used when the number involved is two, whereas *among* and *amongst* are for three or more items. *Amongst* and *among* have the same meaning, but *among* is more common in American English. *Amongst* is a British spelling of *among*.

can/may

Can refers to capability; *may* refers to probability.

Example 13.28. Can/may

We certainly <u>can</u> finish the test, but we <u>may</u> not meet the deadline.

---

cannot/can't/can not

Cannot is one word. Do not use *can not* except for *can not only*. *Can't* is a contraction of *cannot*, and it is meant for informal communication only.

Example 13.29. Cannot/can't/can not

Wrong: Carbon dioxide <u>can not</u> reach its supercritical state if the temperature is below 31.1 °C.

Informal: Carbon dioxide can't reach its supercritical state if the temperature is below 31.1 °C.

Correct: Carbon dioxide <u>cannot</u> reach its supercritical state if the temperature is below 31.1 °C.

Wrong: Your article cannot only report the results; you need to add the discussion part.

Correct: Your article <u>can not only</u> report the results; you need to add the discussion part.

---

compare/contrast

You *compare* to establish similarities or differences or both, and you *contrast* to point out only the differences (not the similarities). When they function as verbs, *compare* is followed by *to*, and *contrast* is followed by *with*. When *contrast* is used as a noun, it is followed by *between*.

Example 13.30. Compare/contrast

- We <u>compared</u> our results <u>to</u> earlier ones. Our results <u>contrast</u> sharply <u>with</u> earlier ones. (*contrast as a verb*)
- There is a sharp <u>contrast between</u> our results and the earlier ones. (*contrast as a noun*)

---

complement/compliment

Complement is a verb, and it means *complete* or *bring to perfection*. It acts as either a noun or a verb. *Compliment* means a polite expression of praise. *Compliment* can be used as a noun or verb.

> **Example 13.31. Complement/compliment**
> - Our results <u>complement</u> the earlier findings reported earlier in the literature.
> - All reviewers <u>compliment</u> the high quality of the work and the manuscript. (*compliment as a verb*)
> - Thanks! I consider that as a <u>compliment.</u> (*compliment as a noun*)

complete/finish

Complete and *finish* are confusing words, especially to non-native English speakers. *Complete* is a positive word, which means that something has no missing parts. *Finish* means *to end* or *stop*. In a study that involves chemical reactions, for example, that the reactions *finish* does not mean the reactions are *complete*. (All reactants are consumed.)

complex/complicated

A *complex* system has many components. Complexity indicates the level of components in the system but does not necessarily evoke difficulty. *Complicated* is used for something causing a high level of difficulty.

consist of/comprise of/compose of

They all mean *be made up* or *be made of*. None of them have gerund phrase (*−ing* form). Their first difference lies in formality: *compose* is more formal than *comprise*, and *comprise* is more formal than *consist*. Second, we only use active voice for *consist of* and only passive voice for *compose of*, but we can use either active or passive voice for *comprise of*.

> **Example 13.32. Consist of/comprise of/compose of**
>
> - This thesis consists of five chapters.
> - This thesis comprises of five chapters.
> - This thesis is comprised of five chapters.
> - This thesis is composed of five chapters.

continual/continuous

Continual means something happens over and over. It repeats many times. *Continuous* implies that something continues without interruption.

> **Example 13.33. Continual/continuous**
>
> - Effective technical writing requires <u>continual</u> learning and practice.
> - The fluids are usually assumed <u>continuous,</u> rather than discrete, in computational fluid dynamics models.

criteria/criterions

Both *criteria* and *criterions* are plural nouns; their singular form is *criterion*, which means *a standard by which something may be judged*. However, the word *criteria* is used in formal writing.

data/datum

In formal writing, *data* is a plural noun, and its singular form is *datum*. In informal writing, *data* can be used as a singular noun.

Example 13.34. Data/datum

Informal: The experimental <u>data</u> in Fig. 6 <u>supports</u> the argument.
Formal: The experimental <u>data</u> in Fig. 6 <u>support</u> the argument.

---

definite/definitive

Definite means known for certain; *definitive* means conclusive. Both imply something precisely defined.

Example 13.35. Definite/definitive

There is a <u>definite</u> correlation between global temperature and the carbon dioxide concentration in the atmosphere; however, it might not be <u>definitive</u> to say global warming is solely caused by CO_2.

---

despite/in spite of/although/though

In spite of and *despite* have a similar meaning to *although* or *though*. Both *despite* and *in spite of* introduce nouns; *although* and *though* introduce subordinate clauses. *Despite* is slightly more formal than *in spite of*, and both emphasize efforts to avoid blames.

Example 13.36. Despite/in spite of/although

Informal: <u>In spite of</u> our efforts, the pilot tests failed.
Formal: <u>Despite</u> our efforts, the pilot tests failed.
Neutral: The pilot tests failed<u>, although</u>/<u>even though</u> we tried our best. (*emphasizes result*)

---

different from/different than

Different from precedes a *noun* or noun form, while *different than* may be followed by a *clause*, to complete an expression.

Example 13.37. Different from/different than

- Our approach to the problem is <u>different from those</u> reported earlier.
- We took an approach that is <u>different than what others did</u> earlier.

---

due to/owing to/because of

Due to, *owing to*, and *because of* all mean *caused by*. However, *due to* modifies the nouns, while *owing to* and *because of* should modify the verbs. *Be due to* means *result from*.

Example 13.38. Due to/owing to/because of

Incorrect: Her promotion is <u>because of</u> her outstanding performance.
Correct: Her promotion is <u>due to</u> her outstanding performance.
Incorrect: She is promoted <u>due to</u> her outstanding performance.
Correct: She is <u>promoted because of</u> her outstanding performance.

---

economic/economical

Economic refers to how money works; *economical* means *not wasteful*.

Example 13.39. Economic/economical

The technology was developed in an <u>economical</u> way, but an <u>economic</u> analysis is necessary before technology transfer.

---

emphasize/emphasis on

Emphasize is a verb, and *emphasis* a noun. *Emphasize* something means *put emphasis on* something. Avoid *emphasize on* something.

everybody/everyone/every one

Everybody and *everyone* are interchangeable, and they are used with persons. They both appear to be singular, but meanings can be plural. *Every one* emphasizes individual.

Example 13.40. Everybody/everyone/every one

- <u>Everyone of</u> the co-authors must have some contributions to the work (*actually meaning <u>all</u>*).
- <u>Every one of</u> the co-authors contributed to the work (*emphasizes individual*).

---

few/a few/fewer; little/a little/less

Few, *a few*, and *fewer* are used with countable nouns; *little*, *a little*, and *less* are used with mass nouns. *Few* or *little* sounds less than *a few* or *a little* and implies negativity.

first/firstly; second/secondly; third/thirdly...

Avoid *firstly* and use *first* in formal writing. The same is true with other ordinal numbers.

flammable/inflammable/nonflammable

Both *flammable* and *inflammable* mean *able to burn* even though they look like opposites. Using *flammable* instead of *inflammable* avoids confusing non-native English speakers. *Nonflammable* is *not flammable*.

imply/infer

Imply means *suggest* or *hint* (*implicitly*); *infer* means *draw a conclusion* from evidence (*explicitly*).

in/into/within

In normally means *inside of* an area or space. *Into* is connected to the movement of something. *Into* typically follows a verb like *go*, *come*, *run*, etc. *Within* stresses that something is not further than a prescribed area or space or not later than a time, but *in* does not indicate emphasis.

Example 13.41. In/into/within

- The sensor is <u>in the device</u>. (*emphasizes status*)
- The sensor is put <u>into the device</u>. (*emphasizes movement*)
- The test will be finished <u>in 20 minutes</u>. (*estimates the time needed*)
- The tests must be finished <u>within 20 minutes</u>. (*sets the time limit*)

---

in order to/to

Use *to* instead of *in order to*, although they have the same meaning. You can occasionally use *in order to* for adjusting the pace of presentation or balancing the sentence length.

inside/inside of

The word *of* is *redundant* in the phrase *inside of*. Use *inside* only in formal writing.

in this paper/study/work/research

The *study* is the *research* or *work* that you did. The *paper* is one way to present your *work* (or the *study*); the *paper* is what you are writing. Do not overuse the word *in this paper* or *in this study* – either one should appear less than three times in an article.

keyword/key word

Keyword is a search-related word either online or in an index. *Key word* means *important word*

lay/lie

You can write *lay something* and *something lies*; *lie* and *lay* are not interchangeable. Their past-tense forms are even more confusing. The past-tense forms of *lay* and *lie* are *laid* and *lay*, respectively. (*Lied* is the past-tense form of *lie* when it means *not telling the truth*.)

Example 13.42. Lay/lie

- We normally <u>lay</u> the filter samples on the bench, and the samples <u>lie</u> in the lab for 24 hours before analysis.
- We <u>laid</u> the filter samples on the bench; they <u>lay</u> in the lab for 24 hours before analysis.

---

like/as

Both *like* and *as* are used for comparison. *Like* is a preposition, which introduces a noun that is not followed by a verb; *as* is a conjunction followed by a verb.

maybe/may be

Maybe means *perhaps* or *possibly* (as adverb) or a mere possibility or probability (as noun). *May be* means *something might*.

on/onto/upon

The relationship between *on* and *onto* is similar to that between *in* and *into*. *On* stresses a position of rest; *onto* implies movement to a position. *Upon* emphasizes movement or condition.

> **Example 13.43. On/upon**
>
> The revision of your manuscript will start <u>upon</u> completion of the additional tests.

only/solely

Only and *solely* are often interchangeable in academic writing, but *solely* emphasize exclusivity. Both should precede the word or phrase that it is intended to modify; otherwise, the idea of the sentence may change.

> **Example 13.44. Only/solely**
>
> - <u>Only</u> we reported the effects of temperature on alkane biofuel *conversion rate.* (*Nobody else reported.*)
> - We <u>only</u> reported the effects of temperature on alkane biofuel conversion rate. (*We did not report other information.*)

percentage/percent/percentile

Percentage is a number that is written out of 100, but it cannot be used with a number. *Percent* is normally used to quantify *percentage*. A *percentile* is a *percentage* of values found below a specific value.

> **Example 13.45. Percentage/percent/percentile**
>
> Two out of 20 students received grades of 150 or lower out of a total of 200 in a test. The <u>percentage</u> of students who received grades of 150 or lower grades is 10 <u>percent</u> (10% calculated from 2/20); the 90th <u>percentile</u> on the test is 150.

precision/preciseness

Precision shows the ability to consistently repeat data or the number of digits indicating the reliability of a measurement. *Preciseness* shows the quality of being accurate, especially about details. *Precision* and *preciseness* are synonyms, but *precision* is more frequently used for measurements than *preciseness* is.

Example 13.46. Precision/preciseness

- You can improve <u>the precision of measurement</u> by changing the meter range from 1–1000 to 10–100 nm.
- You can further improve the quality of your writing by checking <u>the preciseness of punctuation</u>.

principal/principle

Principal can be a noun or an adjective, meaning *main*, *primary*, *most important*, or *influential*, or a person of such importance. *Principle* is a noun only and it means *a rule or code of conduct* or *the basics of truth*.

Example 13.47. Principal/principle

The <u>principal</u> objective of this study is to understand the <u>principle</u> of air conditioning.

raise/rise/arise

Both *raise* and *rise* mean *move up*. *Raise* is a regular transitive verb, followed by a noun (object); *rise* is an irregular intransitive verb preceding a noun (subject). You can write *raise something* and *something rises*. *Arise* mean *to begin to occur* or *to stand up*.

Example 13.48. Raise/rise/arise

When you <u>raise</u> your arm, its gravitational potential energy <u>rises</u>. When you <u>arise</u> from your chair, your potential energy <u>rises</u> too.

reason is/because

Because is a concise alterative of *the reason is that*. However, avoid *the reason is because*, which is redundant.

regarding/with regard to/in regard to

They are interchangeable, meaning *with respect to* or *concerning,* except *regarding* is more concise that the others. However, there are no such phrases as *in regards to* or *with regards to*.

regardless of/irregardless

Regardless of means *without regard* or *considering*. It disqualifies *irregardless* as a formal word, because *irregardless* expresses a double-negative. Use *regardless* instead of *irregardless*.

Example 13.49. Regardless of/irregardless

This paper is technically valuable to readers in the field <u>regardless *of*</u> its informal writing style.

respective/respectively

Respective (adjective) means *belonging* or *relating separately to individuals*. *Respectively* (adverb) means *in the order mentioned*. *Respectively* normally appears at the end of the sentence.

Example 13.50. Respective/Respectively

Wrong: Equations 1–3 can be <u>respectively</u> used for the calculation of A, B, and C.

Wrong: Equations 1–3 can be, <u>respectively,</u> used for the calculation of A, B, and C.

Right: Equations 1–3 can be used for the calculation of, A, B and C, <u>respectively</u>.

Wrong: The temperatures at the top and bottom surfaces of the plate are <u>respectively</u> 550°C and 560°C.

Wrong: The temperatures at the top and bottom surfaces of the plate are, <u>respectively</u>, 550°C and 560°C.

Right: The temperatures at the top and bottom surfaces of the tube are 550°C and 560°C, <u>respectively</u>.

Wrong: The cleaved structure contains four Bi_2WO_6 molecules and five Bi atoms, <u>respectively</u>.

Right: The cleaved structure contains four Bi_2WO_6 molecules and five Bi atoms, <u>~~respectively~~</u>.

Wrong: Specifically, these peaks are assigned to the bending vibrations of Bi-O (445.18 and 576.05 cm^{-1}), WO (727.17 cm^{-1}), and W-O-W (819.61 cm^{-1}) bond, respectively.

Right: Specifically, these peaks are assigned to the bending vibrations of Bi-O (445.18 and 576.05 cm^{-1}), WO (727.17 cm^{-1}), and W-O-W (819.61 cm^{-1}) bond, ~~respectively~~.

---

repeat/replicate/reproduce

Repeat by the same team with the same experimental setup; *replicate* by a different team with the same experimental setup; *reproduce* by a different team with a different experimental setup.

shall/will

In engineering academic writing, you can use *will* and ignore *shall* for future tense. *Shall* has a very strong tone, and it is occasionally used in legal documents.

some time/sometime/sometimes

Avoid *some time*, *sometime*, and *sometimes* in engineering publications, where readers in engineering expect precision and accuracy. *Some time* means *a duration*; *sometime* means *a time to be determined*. *Sometimes* mean *occasionally* at unspecified times. They all are vague and lack preciseness and specifics.

Example 13.51. Some time/sometime/sometimes

- We waited for <u>some time</u> to ensure that chemical reactions stopped.
- Let's get together <u>sometime</u> so we can analyze the data.
- The lab director <u>sometimes</u> comes to the lab to check the safety.

such as/for example/including/etc./and so on/and so forth/and the like

All can be used with *incomplete* lists, and none of them should be used with a complete list. *Such as*, *for example*, or *including* introduces an incomplete list; *etc.*, *and so on*, *and so forth*, or *and the like* ends an incomplete list. However, using any two of these phrases together causes redundancy.

use/utilize/usage

Avoid using *utilize* and *usage* in engineering academic writing because it sounds pretentious. *Use* can be a noun and a verb, depending on context. *Utilize* is a verb. *Usage* is a noun; it means the action of *using something* or the fact of *being used*.

whether/if

Use *if* to introduce a conditional sentence and *whether* to communicate the notion of choice. Use *whether* instead of *whether or not* for conciseness. *As to whether* is redundant.

Example 13.52. Whether

The solvent selectivity depends on <u>whether</u> ~~or not~~ it contains salts. The solvent selectivity is greater <u>if</u> the salt concentration is higher.

---

while/although/whereas

Using *while* often results in ambiguity when it means to substitute *and*, *but*, *although*, or *whereas*. Avoid using *while* for this purpose. Instead, use the *words* that you hope *while* to substitute. Use *while* only when it means *during the time that*.

Example 13.53. While

I wrote this book <u>while</u> the university was closed because of the COVID-19 pandemic.

---

whose/of which

Whose is used with persons; *of which* is used with inanimate objects. *Whose* is occasionally used with an inanimate object to achieve conciseness.

Example 13.54. Whose/of which

This book, the title <u>of which</u> is plain, was written by Dr. Tan, <u>whose</u> native language is not English.

13.9 Practice Problems

Question 1: (Redundancy and wordiness) Rewrite these sentences to improve conciseness.

1. (*Title*) Current Status of Dry SO2 Control Systems
2. In the first enrichment step, WBCs labeled by magnetic beads were depleted by negative IMS. Then the second enrichment step was carried out on using a microchannel for filtration.
3. Morphologies of the samples were characterized by FE-SEM and TEM techniques. It can be seen that 1.5MHNO sample looks like a flower with dewdrops. Specifically, the flower-like base is made up with large nanosheets, which are decorated by small "nanocrumbs." The large nanosheets are composed of Bi_2WO_6, while the nanocrumbs are $WO_3 \cdot 0.33H_2O$. The presence of these two substances is consistent with the XRD pattern of 1.5MHNO (Liu et al. 2020).
4. We collected available field measurements of personal exposure or infiltration factors of ambient air pollutants to compare with modeled results.
5. Table 1 summarizes the specifications of three different types of commercial nanofibrous media tested in this study.
6. Uncertainty in flow rate was calculated based on the larger of the manufacturer's reported uncertainty of ±5% or the standard deviations from three replicate tests.
7. On the other hand, it has been shown that the aerosol type (KCl or diesel particles in size range of 30–1000 nm) significantly affect CADR.
8. In the last decade progress has been made to find a computational alternative to partly replacing physical wind tunnel tests which may be influenced by unpredictable factors (Bai et al. 2013).
9. Model results under various conditions show the separator thickness had a strong impact on the battery energy density.
10. Thus oxidation and absorption of NO by using KMnO4 under weakly alkaline conditions may be a prospective choice for NO control. However, literature survey reveals that the studies on NO absorption using weakly alkaline KMnO4 solution are lacking. In addition, when NO is removed by using oxidation–absorption process, NO2 absorption is an important reaction step (Pan et al. 2012).
11. It is timely to examine the current status of SO_2 scrubbing technologies (Source: doi.org/10.1080/10473289.2001.10464387).
12. There are many substances in the air which may impair the health of plants and animals (including humans) (Source: newworldencyclopedia.org/entry/Air_pollution).
13. In a study, researchers have measured filter penetration and leakage for wide size ranges of particle under different flow rate employing different types of respirators.
14. There are many types of hydrodynamics-based microchannels, including deterministic lateral displacement (DLD), pinched flow fractionation (PFF), inertial focusing [14, 60], etc.

Question 2: (Nominalization) Replace nominalization with verb style.

15. The key feature is that single CTC isolation after capturing process can be achieved by laser cut of the nanofiber substrate.
16. Comparison tests, including three replicate samples for each media type, were conducted to establish the performance of the three media.
17. Although a number of studies have been done both experimentally and theoretically on the effect of loading on filtration efficiency of micron-sized fibers, limited studies were found for this effect on nano-sized fibrous filters.

18. A series of tests exploring different aspects of media efficiency were conducted as shown in Figure 10.
19. Besides particle loading, charge masking or discharge of filters over time may also cause reduction in the filtration efficiency.
20. IPA causes the filtration efficiency of media M2 to be negative for sub-500 nm particles in the chamber test...Overall, IPA caused the filtration efficiency of tested electrospun nanofibrous media to be decreased.
21. In previous work, similar numerical simulations were already conducted with an NACA airfoil (Bai et al. 2013).
22. Model results under various conditions show the separator thickness had a strong impact on the battery energy density.
23. The filtration of CTCs is typically performed using membrane filters.

Question 3: (Precise wording) Revise the following sentences with precise wording.

24. It has been summarized in Section 2.1.3 that DEP suffers from a low loading capacity to ensure its sufficient lateral deflection.
25. In addition, the nanofiber and composite substrates captured 7–15% more MCF-7 cells than microfiber substrate did. This can be attributed to the nanoscale interactions between nanofibers and CTCs.
26. Under the influence of magnetic field, the labeled CTCs experienced lateral migration while unaffected blood cells were swept by buffer flow.
27. The difference in exposure factors of various air pollutants is mostly determined by their infiltration processes, as human activity parameters are the same while modeling different air pollutants. (*Source: Hu* et al. *Environment International (2020) 144: 106018*)
28. The effective loss rate for the PAC was determined as the difference between the rate of contaminant reduction in the test chamber when the PAC is running and the rate of natural decay when the PAC is not running.
29. Uncertainty in flow rate was calculated based on the larger of the manufacturer's reported uncertainty of ±5% or the standard deviations from three replicate tests, whichever is larger.
30. No matter which test we use, the media type is the most important feature we must evaluate.
31. Overall, IPA likely has a complicated impact on the filtration efficiency of nanofibrous media and depends on both the fiber diameter and initial charge states of media.
32. In this paper, a numerical study was conducted to analyze the impact of separator design parameters on the LIB performances.
33. The effective loss rate for the PAC was determined as the difference between the rate of contaminant reduction in the test chamber when the PAC is running and the rate of natural decay when the PAC is not running.
34. Uncertainty in flow rate was calculated based on the larger of the manufacturer's reported uncertainty of ±5% or the standard deviations from three replicate tests, whichever is larger.
35. The G1 meshes used for the rigid, wake and remaining buffer region are, respectively, (594 × 88), (64 × 160), and (104 × 39), where the first and second numbers are the number of hexahedral cells in, respectively, the tangential and radial directions (Bai et al. 2013).
36. However, smaller fines succeed in worming into the downstream catalytic trickle bed. As a matter of fact, even low concentration of fine solids, in the range of several parts per million in the hydrotreater feed, may cause, over months of operation, substantial deposition throughout the catalyst bed. (*Edouard* et al., *2006. Chemical Eng Sci 61: 3875–3884*)

37. In general, the airflow pattern in indoor environments is mostly predominated by air ventilation systems designed and operated for specific requirements. The importance of flow patterns is that it strongly affects droplet's trajectory, which plays a significant role in disease transmission as suggested by others aforementioned.

38. (Comparison of models) The models mentioned above have been compared with respect to their ability to represent experimental sulfation conversion data obtained over a broad range of temperature and SO_2 concentration.

39. According to the above formula, the quality factors of filters with different structures under different particle size ranges are calculated, as shown in Figure 4. (*Zhang* et al, *2021, Building and Environment 187:107392*)

40. Therefore, the main focus of the present paper is to respond to three important issues: (1) is the potential measured when the filter is fixed on a ground plane the same as that in the real case where the filter is left free? (2) Does the presence of the ground plane under the filter affect the potential decay rate? (3) Is the air filtration efficiency affected by the ground electrode too?

41. Despite that the initial potential are not the same, the charge density is the same, and the decay curves must have the same shape too.

42. All of these parameters enable better control of the combustion process, which in turn can allow for lower CO2 emissions, increased fuel economy, and/or more power (EPA 2020).

43. Engine idle stop/start systems allow the engine to turn off when the vehicle is at a stop, automatically restarting the engine when the driver releases the brake and/or applies pressure to the accelerator (EPA 2020).

44. The third option for complying with the CH4 and N2O standards allows manufacturers to propose an alternative, less stringent CH4 and/or N2O standard for any vehicle that may have difficulty meeting the specific standards (EPA 2020).

45. Therefore, since the diameter of the nanoparticle is much smaller than the mean free path of the air, the assumption that the nanoparticle does not interact with other gas molecules is valid.

46. An ideal surface is a flat surface with no asperities, contaminations, and functional groups. Since silver is a high energy material, an ideal silver substrate is a high surface energy (or surface wettability) one.

47. This observation is very important since the characterization studies of the filter media has been carried out so far with filter samples placed near a ground plane while the air filters are left in a free state in industrial applications.

48. Therefore, the discharge of domestic sewage alone increased from approximately 1.75 million tons in 1986 to approximately 29.07 million tons in 2018. Besides, vehicle exhaust emission can also produce a large amount of waste gas containing Pb. According to the statistics, the number of vehicles ownership has been reached 3.367 million by 2018 in Shenzhen, and approximately 6734.5 tons of Pb will be generated.

49. It can be seen that there was a positive correlation between Cu and Ni, Cr and Pb, Cu and Pb, which indicated that Cr, Pb, Ni, and Cu could have synergistic effects on the structure of the microbial community. Besides, there were more than half of bacteria have a negative correlation with HMs, especially for the *Lysobacter, Pseudomonas, Hydrogenophaga, Acinetobacter*, and *Pedobacter* that have a strong negative correlation....

50. In Section 4.1, the ionic concentration distribution, membrane potential, and angular displacement of the main pulvinus are simulated and compared with the results of the published experiments. (*Wang and Li, 2020 Bioelectrochemistry 134:107533*)

51. As shown in Fig. 3d, the normalized deflection $d = d_{max}$ of the beam-like pulvinus is simulated by the model and compared with the published experiment. (*Wang and Li, 2020, Bioelectrochemistry 134:107533*)

52. Various characteristics of turbulent pipe flow are introduced and contrasted to laminar flow. It is shown that the head loss for laminar or turbulent pipe flow can be written in terms of the friction factor (for major losses) and the loss coefficients (for minor losses). (*Source:* Munson et al. 2009)

53. In contrast to the previous findings regarding the mixed aerosol type detection frequency in North America (mixture was rarely detected, based on observations from 97 AERONET stations) [181], the frequency of mixed aerosol detection revealed in this work is much higher. (*Curtsey: M. Munshed*)

54. Use graphics instead of tables. Keep your graphics simple. A graph that is too complicated, containing too much information, will not be easy to read or understand. (*Source:* http://contentliteracytraining.pbworks.com/f/WritingthatWorkscontentteachers.pdf)

55. Addition of makeup sorbent is not simple, and movement requires the use of bucket elevators. A complicated regeneration sequence is necessary to control temperature in the regenerator. (*Vamvuka* et al. *Environmental Engineering Science (2004) 21: 525–548*)

56. Since this is an introductory text, we have designed the presentation of material to allow for the gradual development of student confidence in fluid problem solving. Each important concept or notion is considered in terms of simple and easy-to-understand circumstances before more complicated features are introduced (Munson et al. 2009).

57. The Bernoulli equation is introduced in Chapter 3 to draw attention, early on, to some of the interesting effects of fluid motion on the distribution of pressure in a flow field. We believe that this timely consideration of elementary fluid dynamics increases student enthusiasm for the more complicated material that follows (Munson et al. 2009).

58. The flow field may be quite simple as in the above one-dimensional flow considerations, or it may involve a quite complex, unsteady, three-dimensional situation such as the flow through a human heart as illustrated by the figure in the margin (Munson et al. 2009).

59. Since model year 1975, the technology used in vehicles has continually evolved. Today's vehicles utilize an increasingly wide array of technological solutions developed by the automotive industry to improve vehicle attributes discussed previously in this report, including CO2 emissions, fuel economy, vehicle power, and acceleration (EPA 2020).

60. Transmissions that have seven or more speeds, and continuously variable transmissions (CVTs), allow the engine to more frequently operate near its peak efficiency, providing more efficient average engine operation and a reduction in fuel usage.

61. The gas stream passed through the stainless steel filter while an online flue gas analyzer (Green Line eurotron) equipped with an SO_2 sensor continuously monitored the SO2 concentrations of the outlet stream. (*Tseng* et al.*, 2011, J of A&WMA 52 (11): 1281–1287*)

62. (*Source:* https://www.researchsquare.com/article/rs-34150/v1) Criteria for selection of the houses is made on the basis of the following criteria adapted from
    (a) Functional homestead (at least mother and children present)
    (b) Cook daily in designated cooking area in home
    (c) Use firewood as main fuel on a daily basis
    (d) Firewood is either bought or collected in fixed bundles on a regular basis
    (e) Involvement in the kitchen work

63. Studies in Asia (Huang et al. 2015; Lai et al. 2015; Li et al. 2019; Shen et al. 2017; Xu et al. 2014; Xu et al. 2016), North America 61 (Evangelopoulos et al. 2020; MacNeill et al. 2014; Wallace and Williams 2005), and Europe (Meier et al. 2015; Zauli-Sajani et al. 2018) show that infiltration and exposure factors vary significantly among various air pollutants, regions, seasons, and populations due to variations in the pollutant and architectural characteristics; ventilation; and human activities. (*Source: Hu* et al. *Environment International (2020) 144: 106018*)

64. Nevertheless, although the economic value of CAES having multiple stream revenues has been studied in the context of planning and operation of power systems, the actual impact of CAES facilities on the electrical grids have not been properly addressed in the literature, in part due to the lack of suitable models.

65. The continuing decrease in pickup truck 0-to-60 times is likely due to their increasing power, as shown in Figure 3.8. While much of that power is intended to increase towing and hauling capacity, it also decreases 0-to-60 times (EPA 2020).

66. After screening various dienes, the effect of different substituted alkynes was investigated. Firstly, two dialkyl acetylenedicarboxylates were employed in the cycloaddition and the results implied that larger steric hindrance of the acetylenedicarboxylate would result in lower reactivity (Table 2, 4s and 4t).

67. Until recently, researchers tried to build physical models of spatially variable soil which were used for laboratory tests. Chakrabortty et al. (2008) firstly constructed deposit models of heterogeneous sand to conduct centrifuge liquefaction tests of heterogeneous soil successfully.

68. The results demonstrate two things. Firstly, the data collected were in favor of sustainable practices of purifying the air. Second, their willingness to investing in the said practices.

69. Based on measurements performed in 1998, the upper limit of the plume conversion rate of SO_2 to SO_4^{2-} was found to be $4.4 \pm 3.6\%$ hr^{-1}, consistent with the dryrates found for the CUF plume under similar conditions in 1978 before installation of FGD wet scrubbers. This implies that the benefit of SO_2 emissions reductions on amounts of secondary SO_4^{2-} formed in fine particles will likely be realized (Imhoff et al. 2000).

70. The other peaks in the NCO-HS/LiPS shift towards lower BE, which implies the coordination between the oxygen defects and the sulfur in LiPS, corresponding to the PS adsorption on oxygen vacancies.

71. The STEM image and corresponding EDX element mapping in Figure 5.1c implies a homogeneous element distribution of Co, S, and N.

72. If the ion signal exceeded a threshold value within a set time period after the laser was fired, a particle was considered to have been "hit" and the mass spectrum was saved. (*Kane* et al. *Anal. Chem. (2002) 74: 2092–2096*)

73. The peak area of the m/z 55 ion from oleic acid remains constant within experimental error between 200 nm and 80 nm, while the detection probability drops by a factor of six. (*Kane* et al. *Anal. Chem. (2002) 74: 2092–2096*)

74. For these cases stop distance was proportional to particle diameter within the limits of experimental error and uncertainty in particle size. (*Dahneke and Friedlander 1970, J of Aerosol Sci. 1:325–339*)

75. When a power plant plume is sampled far from the source, it has been diluted thousands of times with air from the surroundings. In order to determine the contribution of the source to levels in the plume, the mass contributed by the background air must be subtracted. Because of the high levels of dilution, it is critical that the background be accurately characterized (Imhoff et al. 2000).

76. For example, the average fuel economy for a set of 10 vehicles, with three 30 mpg vehicles, four 25 mpg vehicles, and three 20 mpg vehicles would be (note that, in order to maintain the concept of averaging, the total number of vehicles in the numerator of the equation must equal the sum of the individual numerators in the denominator of the equation) (EPA 2020).

77. In order to bracket the possible GHG emissions impact, Table E.4 provides ranges with the low end of the range corresponding to the California power plant GHG emissions factor, the middle of the range represented by the national average power plant GHG emissions factor, and the upper end of the range corresponding to the power plant GHG emissions factor for part of the Midwest (Illinois and Missouri) (EPA 2020).

78. Measurements are carried out in the condition that the number concentration out of the door is lower than inside. If the number concentration outside the door coming from other dental offices is higher than inside, particles may enter the room and worsen the inside air quality.

79. Valve timing has evolved from fixed timing to variable valve timing (VVT), which can allow for much more precise control. In addition, the number of valves per cylinder has generally increased, again offering more control of air and exhaust flows. All of these changes have led to modern engines with much more precise control of the combustion process (EPA 2020).

80. Get in the habit of using a consistent system of units throughout a given solution. It really makes no difference which system you use as long as you are consistent; for example, don't mix slugs and newtons (Munson et al. 2009).

81. These results indicate that both the drag and lift as predicted by potential theory for a fixed cylinder in a uniform stream are zero. Since the pressure distribution is symmetrical around the cylinder, this is not really a surprising result. However, we know from experience that there is a significant drag developed on a cylinder when it is placed in a moving fluid (Munson et al. 2009).

82. Actually, Eq. 1.2 is a special form of the well-known equation from physics for freely falling bodies (Munson et al. 2009).

83. When vapor bubbles are formed in a flowing fluid, they are swept along into regions of higher pressure where they suddenly collapse with sufficient intensity to actually cause structural damage.

84. Actually, the reduction mechanism of sulfur is much more complex in nature compared to that of metal oxides and remains elusive and debated among the research community.

85. The gas–solids two-phase contacting operations can be classified to several kinds of flow regimes, such as fixed bed, bubbling fluidization, slugging fluidization, fast fluidization and turbulent fluidization, etc., each of the regimes occurs over a definite range of gas velocities and gas and solid properties.

86. In a bubble column, the value of kLa is associated with numerous parameters such as gas superficial velocity, gas distributor design, bubble column diameter, etc. (*Source: Yu H. 2012, PhD Thesis, Waterloo, Canada*)

87. If, together with your accepted article, you submit usable color figures then Elsevier will ensure, at no additional charge, that these figures will appear in color online (e.g., ScienceDirect and other sites) regardless of whether or not these illustrations are reproduced in color in the printed version.

88. However, considering significant variability in conductivities, found by us for each set of the devices and also reported elsewhere, it is difficult to conclude whether or not hydrazine reduced IGO flakes are indeed more conductive than their HGO and HGO+ counterparts.

89. We found that it was important to carefully position the injection tube exit with respect to the induction coils: if the tube exit was too high it was difficult to operate the plasma stably because the characteristic recirculation vortex (see Fig. 7) was disturbed, while if the tube exit was too far downstream it was difficult to vaporize all of the injected powder. (*Girshick* et al. *Journal of Aerosol Science (1993) 24: 367–382*)

90. The resulting CO_2-concentrated gas stream that is released in the stripper is recovered, while the resulting CO_2-lean solvent is circulated back to the absorption tower. (*Source:* hdl.handle. net/2142/16828)

91. While one-dimensional models provide some insight into the process, real plasma reactors are far from being one-dimensional, because the walls must be cooled to avoid melting, resulting in steep radial temperature gradients. (*Girshick* et al. *J. of Aerosol Science (1993) 24: 367–382*)

92. While positive ion mass spectra are normally used to characterize ultrafine particles (defined here as particles smaller than 200 nm in diameter), negative ion mass spectra can provide complementary information. (*Kane* et al. *Anal. Chem. (2002) 74: 2092–2096*)

93. The absence of water reduces the order of the reaction to 0.76 in respect to SO_2, while in the presence of water vapor the reaction becomes first order. (*Vamvuka* et al. *Environmental Eng Sci (2004) 21: 525–548*)

94. IMS is a fairly mature method for CTC enrichment which has received considerable investigations.

95. Due to the amplified role of surface energy in determining the internal energy of the nanoparticle and top lattice layers of the substrate, it is essential to construct and simulation the changes of their lattices and geometries from the first-principle approaches.

96. At the bottom of CB, the addition of C 2p and a small amount of 2S orbitals leads to a slight reduction of the band gap, and makes it possible for the e at the top of the VB to transfer to the empty C 2p orbitals.

Visuals

14

14.1 Using Simple Visuals

It may be a *cliché* to say "A picture is worth a thousand words," but visuals do have power. Effective communication does require a multisensory experience. Readers notice visuals first when the visuals are placed beside sentences and paragraphs. In addition, a nice illustration in the crowd of words gives the readers a visual relief.

Technically, visuals convey messages more effectively than text alone. For example, a schematic diagram can show how a complex system works, which may take multiple paragraphs or pages to explain. Charts and tables can also describe the relationship between numbers more clearly than words do. You are encouraged to use appropriate visuals to enhance your writing.

Like with text, visuals also should be created and presented with clarity, conciseness, and consistency. Visuals and the related text most often support each other. The visuals aid readers in understanding the text, but they cannot replace the text. Meanwhile, the contents of the visuals appear in the text that explains the visuals. You can follow these general guidelines on visuals when you revise your draft.

Visuals frequently used in engineering academic writing include figures and tables, where figures can be charts, schematics, graphs, images, maps, and so on. You can create and process numerous types of visuals using software like *MS Office*, but using these tools is beyond the scope of this book. This chapter is focused on some key factors that are more important to engineering writing than other types of writing.

> **Note**
>
> Use copyrighted materials only with written permissions from the copyright holders. Reproduced works also require permissions.

14.2   Key Elements in Visuals

When you create visuals, include only essential information that supports your ideas of writing. You can place lengthy, detailed visuals into appendices or supplemental materials if they are nonessential to the body of the text. In short, you can use the following techniques to avoid visual clutter.

- Use text with typeface and font compatible with the main text.
- Use symbols globally known (e.g., % for percentage; °C for degree Celsius); otherwise, spell out the words if the symbols are defined for your document only.
- Specify the scales and the units of measurement (e.g., axis's for line graphs).
- Use terms consistent with the main body.

14.2.1 Captions of Visuals

Captions are titles that describe their visuals. The function of a caption to a visual is like that of a title to the main document. The caption should capture key messages in the visual. The figure caption normally appears below the figure, but the table caption appears above the table.

You need to separately number the figures and the tables. They follow two different orders of sequence. Their numbers and captions are referred to in the List of Tables and the List of Figures, if applicable. (*See also* Table 14.1 for cross-reference of visuals.)

14.2.2 Labels in Visuals

Labels are necessary to both the clarity and the conciseness of the visuals. In schematics, for example, you need to label the parts or components that you explain in the text. In a map, as another example, you need to label the key streets and the locations that are relevant to your research.

You may put labels inside simple visuals with clear and consistent texts. Meanwhile, you need to make sure the labels do not clutter the visuals. In this case, you can label parts of a complex diagram with numbers, and list their texts at the bottom, below the caption, or on the side of the schematic diagram. Regardless of their locations, using compatible font size helps avoid mismatch in size. Fig. 14.1 shows an example with labels integrated with the diagram (Givehchi 2015).

In contrast, Fig. 14.2 has no label. As a result, this figure is useless because readers cannot understand its purpose. The figure is meant to point the readers to the shaded rectangle, which represents the spot for air sampling. In addition, this figure lacks emphasis because of unnecessary lines and shapes.

Table 14.1 Cross-references of visuals and equations

Type	Cross-reference	Meaning
Figure	Figure 6	The sixth figure
	Figure 6	Same as Figure 6
	Figure 6-8!	The eighth figure in Chapter 6
	Figure 6-8!	Same as Figure 6-8!
	Figures 6-8!	Figures 6, 7, and 8
	Figures 6-8!	Same as Figures 6-8!
Table	Table 6	The sixth table
	Table 6-8!	The eighth table in Chapter 6
	Tables 6-8!	Tables 6, 7, and 8

Fig. 14.1 Schematic of experimental setup. (Reprinted with permission from R. Givehchi©2015)

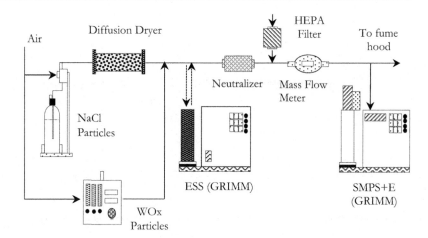

Fig. 14.2 Blurry figure without label. (Reprinted with permission from X. Cheng©2006)

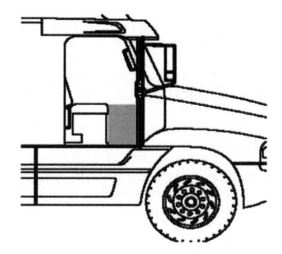

14.2.3 Text in Visuals

Figure 14.1 horizontally positions the explanatory text or labels, from left to right. You should do so whenever possible. This does not apply to the y-axis of a line graph for data presentation. Otherwise, it looks awkward.

Font size mismatch is a typical mistake engineers make. Fig. 14.3 shows a figure of an article that is published in a prestigious international journal (Zhou et al. 2011). This figure has both pros and cons. On the one hand, the figure is produced with clarity: lines, shapes, and texts are properly laid out. On the other hand, it is not clear what the symbols stand for, although there is enough room for explanatory words. Another problem is the font size mismatch between the body text and the caption title of the figure. This size mismatch can be easily fixed, should the authors follow the basic guidelines about visual formatting.

Fig. 14.3 Figure with
mismatched text size.
(copyright © American
Association for Aerosol
Research, reprinted by
permission of Taylor &
Francis Ltd on behalf of
American Association
for Aerosol Research.)

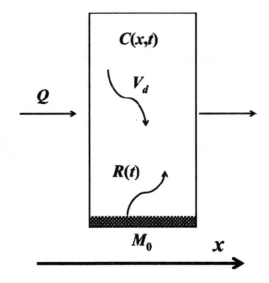

FIG. 1.   Illustration of the process involving airflow, particle resuspension, and
deposition in a duct unit.

14.3   Images

Images are photos processed with software. The original works could be a photo or numerical simulation. Fig. 14.4 shows two images for the morphology of nanofibers, and Fig. 14.5 shows numerical simulation of temperature profile in a battery. Beside high definition, reference scale and text direction are also important for the clarity of these images.

You must use reference scales when size and numerical values matter to your results. In addition, you must avoid wrong scales that may cause visual distortion because they lead to inaccurate presentation.

14.4   Graphs

14.4.1 Line Graph

A graph presents numerical data; it is more apparent and comprehensible than a table for the same dataset. However, graphs may not be as precise as tables; data in a table may contain multiple digits after decimal. You can create line graphs, bar graphs, pie graphs, etc. using data that can be presented in tables.

Engineering publications frequently use line graphs (combined with dots as needed), and other types of graphs are primarily for nontechnical presentations. Fig. 14.6 shows an example of line graph. A line graph allows the comparison between multiple datasets (y-axis) against one common variable (x-axis).

Using appropriate scales improves the clarity of visuals. For example, a logarithmic scale enables displaying data over a wide range of measurement values in a compact way. Logarithmic scales may also reduce misleading information in line graphs.

Fig. 14.4 Processed photos as visual. (Reprinted with permission from Elsevier©2019)

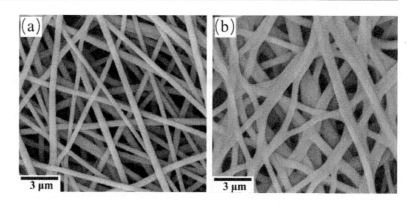

Fig. 14.5 Images for numerical simulation. (Reprinted with permission from ACS ©2020)

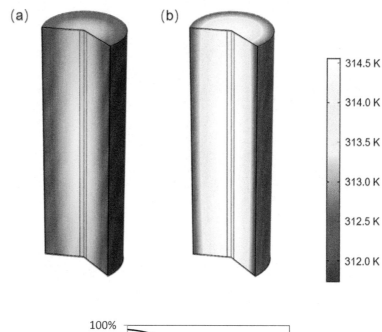

Fig. 14.6 A line graph (RH: relative humidity)

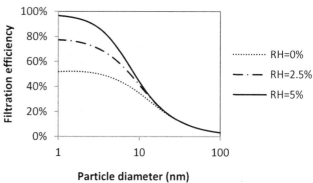

14.4.2 Dot Graph

Figure 14.7 shows a dotted graph (Li et al. 2019). Pay attention to the layout, axis, and text direction in this figure. They are important to the clarity of presentations as well as the visual comfort.

14.4.3 Line-Dot Graph

Figure 14.8 shows that you can use lines and dots in the same graph to compare experimental data with numerical simulations. By default, lines are reserved for the numerical simulation, which can be continuous, and dots are for the experimental data, which are always discrete. Otherwise, the figure confuses readers who expect so.

Fig. 14.7 Dotted graph. (Printed with permission from Elsevier ©2019)

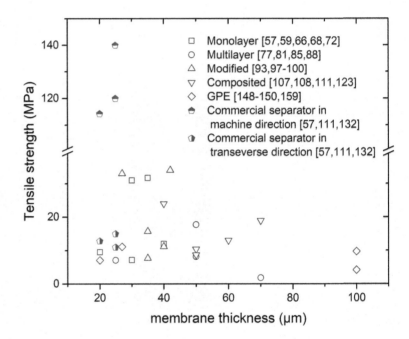

Fig. 14.8 Line-dot graph for comparison between experiment (dot) and simulation (line)

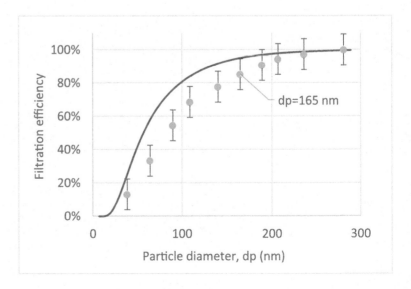

14.4.4 Error Bars

Error bars applied to graphs indicate the precision of measurement. The shorter the error bars, the more repeatable the data are. Error bars are usually based on *standard deviation* or confidence intervals; they can also be the minimum and maximum values in a ranged dataset.

Error bars provide additional details on the results. They affect the accuracy of your results and conclusion. For example, Fig. 14.8 compares a model with experimental data. The averaged experimental results (circles) seem to indicate that the model efficiency is always higher than experimental measurement. However, you cannot draw this conclusion when the error bars are considered. The preceding conclusion is valid only for particles smaller than 165 nm in diameter.

14.4.5 Schematic Diagram

A schematic, or schematic diagram, is a type of visual for illustration of the relationship among components of a system. A schematic is created using graphic symbols and lines instead of real photos. Schematics focus on the emphasis, detail, and relationship that photos cannot illustrate.

Schematics can be simple or complex. They are often used in engineering publications to depict complex systems or processes. Fig. 14.1 shows the schematic of experimental setup used for air filtration studies. Blocks combined with text in the diagram represent major devices and accessories, which are connected by lines. Meanwhile, arrows illustrate the direction of air flow.

Figure 14.9 shows the schematic of a home-made corona charger. It uses only black lines and shapes, and the variation in lines and shapes improves clarity (Givehchi 2015). Shaded housing stands out from the rest because of the contrast effect. Physical parts (isolator, needle, etc.) and model parameters (gap and diameter) are labeled to assist engineering analysis. Labels are aligned on the right with consistence, and white spaces purposely surrounding the elements further enhance visual comfort. Overall, this is a well-prepared visual.

14.4.6 Exploded View

Exploded view is another type of schematic. An exploded view reveals the proper sequence of assembly or the details of individual parts. It also illustrates internal parts and their relationship to the whole device or system. However, exposed view is rarely used in engineering academic publications. They often appear in manuals.

14.4.7 Maps

Maps are useful to the presentation of geographic information. Depending on the purpose of writing, a map may have the following information: climate pattern, population distribution, street direction, site location, traffic, and so forth.

Follow the general guidelines about visuals for the creation and integration of maps into the text. In addition, you need to include a scale for the indication of distance and the direction of north for orientation. You can emphasize key items using color, shade, and so forth. On the other hand, you should remove unnecessary information to avoid visual clutter.

Fig. 14.9 Schematic of a
corona charger.
(Reproduced with
permission from
R. Givehchi@2015)

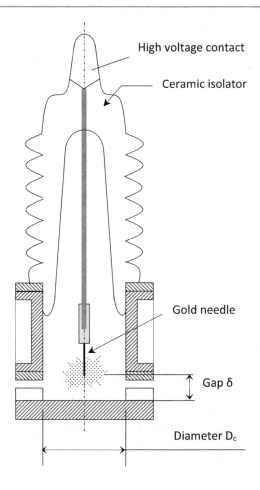

High voltage contact

Ceramic isolator

Gold needle

Gap δ

Diameter D_c

14.5   Tables

Tables are more effective than graphs for the *precise* presentation of data used to support your reasoning and argument. Tables should be descriptive but concise. You can use abbreviations and symbols in
box heads of complex tables. Additionally, footnotes can also define abbreviations or symbols if they
are not defined in the text of document. Within the table body, align the data in adjacent rows or col-
umns for the ease of comparison.

   Always attempt to fit a table on one single page. Otherwise, repeat the table number of the same
table on the new page. The caption title usually does not repeat on the new page. If it repeats, the caption title should be identical to that on the previous page, but it is followed by a comma mark and a
word *continued* or by the word *continued* enclosed in parentheses. For example:

- Table 1. Continued
- Table 1. (continued)
- Table 1. Caption title, continued
- Table 1. Caption title (continued)

14.6   Color

14.6.1 Black-and-White Visuals

Black-and-white illustrations work well for most engineering academic publications. You can differentiate them by line style, grayness, thickness, and dots. Therefore, you use color only when necessary. Fig. 14.6 shows a line graph that is produced using black elements only. It achieves clarity without colors, and different line thickness and styles effectively distinguish the results from each other. Fig. 14.1 uses black only but clear enough.

14.6.2 Color in Visuals

You may use color for highlighting or emphasis in figures and tables; however, you should use color with care. Choose neutral colors for the figures when you need colors.

Misused color can cause frustration in reading, especially a sharp color like red. You should use sharp color only when necessary, although it enhances contrast and emphasis. Particularly, red should be used only when it is necessary. For example, a red cross is often used as a medical sign in North America, but red represents danger and a green crescent signifies medical services in Muslim countries. On the contrary, red is generally a positive color in east Asia, especially the greater China areas. However, a red cross marked over an individual's name means that person is executed.

Figure 14.10, for example, is part of a figure used in a dissertation that is written by a non-native English speaker. The student submitted his dissertation as part of the requirements of doctoral degree at a Canadian university. The dissertation has many visuals produced with unnecessary red color.

Thoughtful writers think beyond their own cultures and experiences when they are writing for the international readers. Make sure that the graphics are free of religious or gender implications. If a human body or body parts are used in the visual, for example, it is recommended to represent the human beings with outlines or neutral abstractions – nudity is eccentric.

Again, black-and-white illustrations work well for most technical writing. When you need color, using consistent color theme is important to professional appearance. Smooth transition from one color to another also ensures comfort reading. On the contrary, color mismatch creates negative visual effects, and it may call for negative perceptions on the creator's capability. For example, two different colors may appear to be the same on a hardcopy printed out in black only.

Figure 14.11 shows a line-bar graph with unneeded color and other errors. Color in this figure is unneeded, and it can be fixed by using black lines and bars. In addition, Fig. 14.11 could further improve with horizontal direction of text for the x-axis and by adding a boarder to enclose the plot area. Finally, replacing the three-dimensional bars with two-dimensional bars or simply black dots would be my personal preference.

Fig. 14.10 Part of a figure
with unneeded color.
(Reprinted with permission
from the creator; anonymous
for moral right)

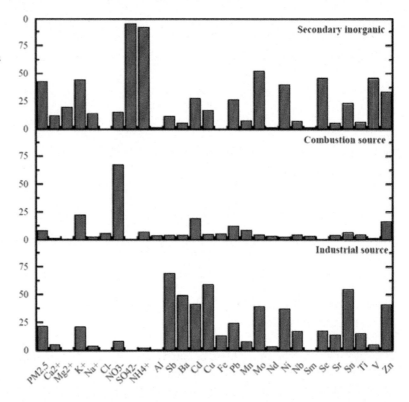

Fig. 14.11 Figure with
unneeded color.
(Reprinted with
permission from the
creator; anonymous for
moral right)

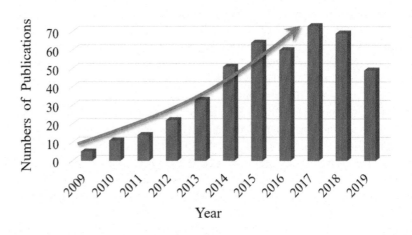

14.7   Integrating Visuals into Text

There are several best practices in integrating visuals into text. In text, you can refer to the visuals by
their figure or table numbers. While doing so, capitalize the word *figure* or *table* (i.e., Figure or Table)
because it cross-references a specific section of the document. Depending on the format guideline,
you can spell out the words of visuals or use their abbreviations. (Same rules apply to the cross-reference of equations, although equations are not normally considered as visuals.) Table 14.1 summarizes
typical spellings of cross-reference of figures and tables.

14.8   Placement of Visuals

Visuals often appear in the main body of text right after their first explanations. Visuals normally follow their first mention in a manuscript under review and a graduate thesis. However, it may change in the final version of journal articles.

14.9   Practice Problems

Sources: All visuals are reprinted with permissions from original creators – anonymous for moral right.

Question 1 (Tables) The following visuals are extracted from draft theses under review. List the places that need refinement.

Table 2 Benzene Toxicity Values

CAS 71-43-2	Oral Exposure	Inhalation exposure
Quantitative Estimate	$1.50*10^{-2}$ per mg/kg-day	$2.20*10^{-6}$ µg/m^3
Test Species	Humans	
Extrapolation Method	Linear extrapolation of human occupational data	Low-dose linearity utilizing maximum likelihood estimates
Tumor Site	Hematologic	
Tumor Type	Leukemia	
Weight of Evidence for Cancer Characterization	Known human carcinogen	
Critical Effect Systems	Bone marrow (Immunotoxicity)	
Reference	U.S. EPA IRIS	

Table 5 Gaps in the existing literature

#	Gap	Addressed in Proposed Framework
1	**Roadway Geometry Characterization**	Yes
2	**Source Characterization**	Yes
3	**Low Temporal Resolution in Mobile Sources Emission Inventory**	Yes
4	**A Disconnect Between EI10, ADM11, and Risk Models**	Yes
5	**MSATs Lack Air Quality Standards**	Yes
6	**Variability And Uncertainty in Risk Results**	Yes
7	**Estimation of Ecological Risk from Exposure to Mobile Source Air Toxics**	Yes
8	**Definition of Ecological Risk Problem Formulation**	Yes

Table 2.2 Surface properties associated to wettability

	Angle	Degree of wettability	Adhesive to cohesive interaction ratio
A	$\theta < 10°$	Superhydrophilic	Very strong
B	$10° < \theta < 90°$	Hydrophilic	Strong
C	$90° < \theta < 150°$	hydrophobic	Weak
D	$150° < \theta < 180°$	Superhydrophobic	Very weak

Table 2.1 Parameters affecting the coefficient of restitution of a nanoparticle colliding on a substrate

Dynamical characteristics of the particle	Impact velocity
	Initial angular velocity
	Angle of incidence
Physical and chemical properties of the particle and substrate	Particle size
	Particle material
	Particle shape
	Surface energy
	Chemical reactivity
Ambient conditions	Temperature
	Humidity

Question 2 **(Figures)** Suggest the revisions needed in the following figures.

1. This figure is part of a figure used in a dissertation that is submitted dissertation as part of the requirements of doctoral of philosophy degree at a Canadian university.

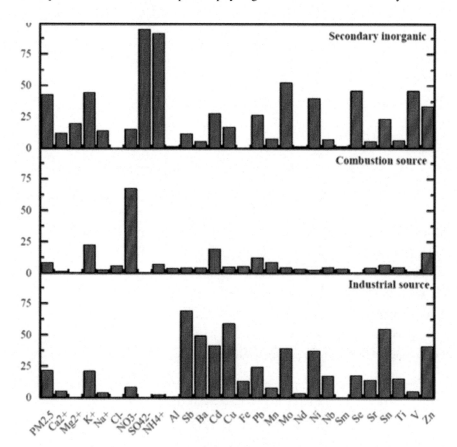

2. Identify the errors and suggest revisions for Figs. 4.3 and 4.7 in the Master of Science Thesis (L. Cheng 2010, *Lignin Degradation and Dilute Acid Pretreatment for Cellulosic Alcohols,* UNIVERSITY OF CINCINNATI, USA.) Available online: https://etd.ohiolink.edu/apexprod/rws_etd/send_file/send?accession=ucin1282329715&disposition=inline

3. Contact angle (θ) of a water droplet on a surface

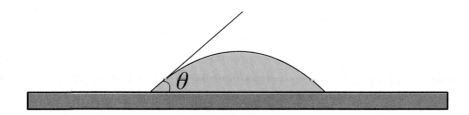

4. Schematic representation of surfaces with different wettability

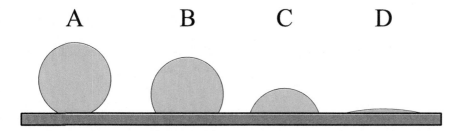

5. Schematic representation of the simulation framework to model a collision process

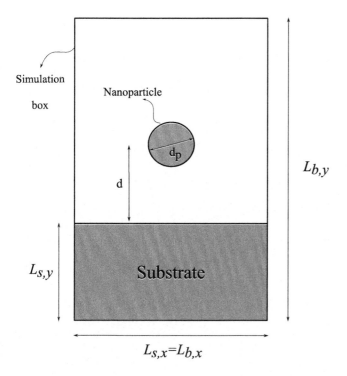

6. Coefficient of restitution versus collision velocity for two colliding nanoparticles. The red circles show the results reported by Takato et al. and the blue curve shows the results from MD simulations.

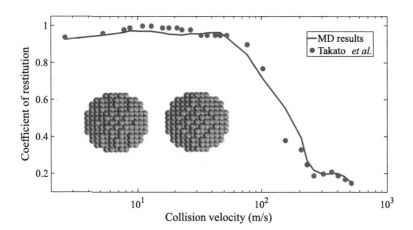

7. Water density profile as a function of temperature. TIP4P-2005 is used to model water molecules. The results from MD simulations (blue line) are compared to the results reported.

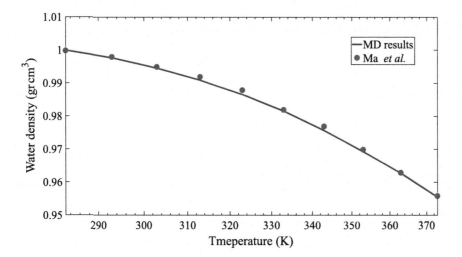

8. Coefficient of restitution of the nanoparticle colliding on the high surface energy substrate as a function of the water layer thickness at different impact velocities.

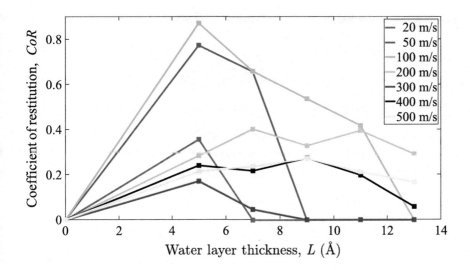

9. Flowchart of the framework

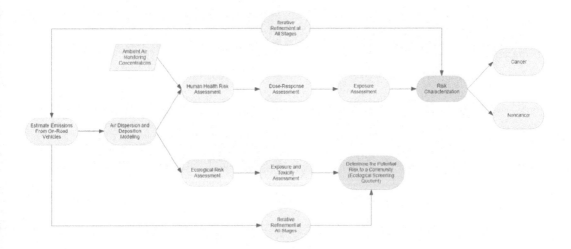

10. Overall research plan of graphene-based material synthesis and application

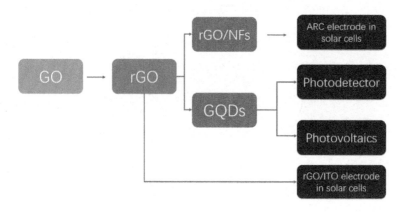

11. (a) HRTEM image of GO, (b) selected area diffraction (SAED) pattern, and (c) Raman spectrum of GO

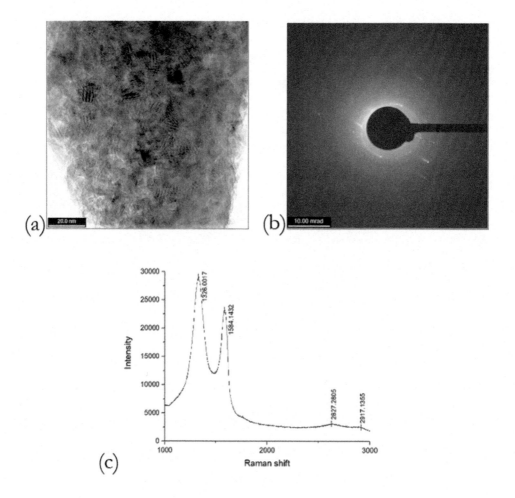

12. FTIR spectra of GO and rGO

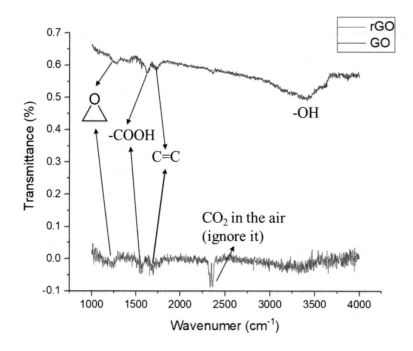

13. Comparison between redox active battery separator layer and (a) PPy particles loaded on inert nanofiber web [14] and (b) free-standing PPy nanofibers (this proposal)

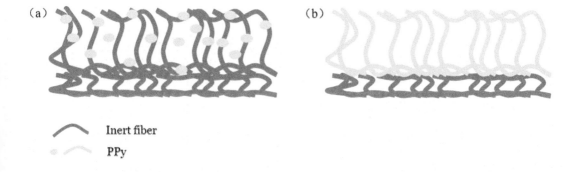

14. Electrospinning setup (a) and coaxial needle (b) used in this study

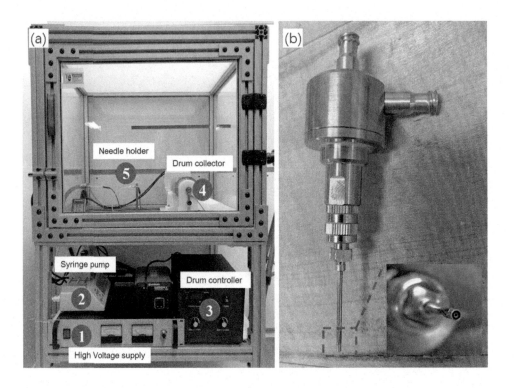

15. FTIR spectrum for samples before and after polymerization

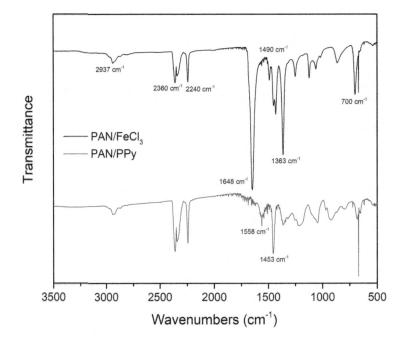

16. Effects of calcination temperature on the activity of HT-C-R catalyst in glucose isomerization at 110°C (initial glucose concentration of 100 mg/ml, feeding flow rate of 0.5 ml/min).

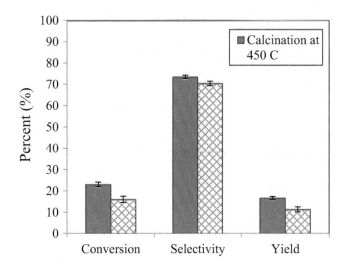

17. TGA graphs for the fresh and used MgO

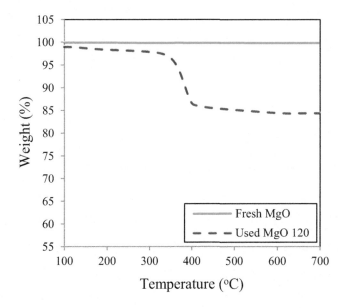

18. Suggest revisions to Figs. 1, 4, and 5 in the sample paper:

- **Sample paper:** M. Yanilmaz, Y. Lu, J. Zhu, X. Zhang, 2016. Silica/polyacrylonitrile hybrid nanofiber membrane separators via sol-gel and electrospinning techniques for lithium-ion batteries, *Journal of Power Sources* 313: 205–212. Available online: https://doi.org/10.1016/j.jpowsour.2016.02.089

Punctuation

15

15.1 Punctuation Marks

Punctuation is an important part of writing. Understanding punctuation is essential to writing with clarity and preciseness. Table 15.1 lists 12 punctuation marks commonly used in academic technical writing in English. They are presented in this chapter following an alphabetic order instead of the order of importance.

15.2 Apostrophe

You can use an apostrophe (') to show possession of nouns (e.g., the *paper's title*; the *authors' addresses*), to form plurals of lowercase letters and abbreviations (e.g., *three y's*), or to indicate contractions (e.g., *don't, int'l*). However, you do not need to use apostrophes with a plural on capital letters or numbers such as *As*, *PDFs*, and *Drs*. In addition, you should avoid contractions in formal writing.

Example 15.1 Apostrophes

- You <u>don't</u> need to write down all <u>authors'</u> contact information. (*Note: avoid contractions in formal writing.*)
- He wrote <u>three *i's*</u> in the same equations with different meanings.
- Replacing all <u>e.g.'s</u> with <u>*for example*'s</u> would not help achieve conciseness.
- The best student in class received four *As* and two *Bs* in his courses.
- This book was written in early <u>1990s</u> by two <u>PDFs</u>.

© The Author(s), under exclusive license to Springer Nature Switzerland AG 2022
Z. Tan, *Academic Writing for Engineering Publications*,
https://doi.org/10.1007/978-3-030-99364-1_15

Table 15.1 Names and marks of punctuation

Name	Mark	Name	Mark
Apostrophe	'	Parentheses	( )
Brackets	[ ]	Period	.
Colon	:	Question mark	?
Comma	,	Quotation marks	" "
Dash	--	Semicolon	;
Hyphen	-	Slash	/

15.3   Colon

Colons (:) are used in sentences, numbers, times, etc. for a variety of functions. A colon acts as a pause and may catch the readers' attention prior to introducing what follows. It can also indicate time or a numerical ratio.

Don't insert a colon between a verb (or a preposition) and its objects. Don't follow *including, such as, for example*, etc. with a colon. A non-native English writer might think it emphasizes what follows. Instead, it becomes a grammatical error.

Don't capitalize the word following colons unless it is a full sentence after the colon mark.

Example 15.2 Colons

- An engineering research project should meet two criteria: the scope of research is new to the field, and the contributions are important to the society.
- A graduate thesis consists of at least five parts: literature review, objectives, methodology, results and discussion, and conclusions.
- Figure 5 shows the correlation between traffic and low air quality between 7:30 a.m. and 9:30 a.m.
- Air is a mixture of nitrogen, oxygen, and other gases at a volumetric ratio of 78:21:1.

Example 15.3 Don'ts for colons

Wrong: Typical catalysts include: enzymes, acid-base catalysts, and surface catalysts.
Right: Typical catalysts include enzymes, acid-base catalysts, and surface catalysts.
Wrong: This manuscript may be submitted to: *Science, Nature,* or *Nature Nanotechnology.*
Right: This manuscript may be submitted to *Science, Nature,* or *Nature Nanotechnology.*
Wrong: This article can be submitted to many prestigious journals such as: *Science, Nature, and Nature Nanotechnology.*
Right: This article can be submitted to many prestigious journals such as *Science, Nature, and Nature Nanotechnology.*

15.4 Comma

Commas and *periods* are the two most frequently used punctuation marks. A comma indicates a brief pause before a sentence ends, and a period ends the sentence. Commas have the following functions in sentences: linking independent clauses, introducing elements, enclosing elements, quotations, separating items, clarifying and contrasting meanings, indicating omissions, etc.

15.4.1 Linking Independent Clauses

A comma mark immediately precedes a *coordinating conjunction* that links an independent clause when both clauses have subjects. However, a comma is unneeded when both clauses share the same subject.

Example 15.4 Comma linking independent clauses

Incorrect: Your description of procedure is clear <u>but you</u> omit the specifications of the devices.

Correct: Your description of procedure is clear<u>, but you</u> omit the specifications of the devices.

Incorrect: Your description of procedure is <u>clear and you</u> can start testing now.

Correct: Your description of procedure is clear<u>, and you</u> can start testing now.

Incorrect: You have a clear description of <u>procedure, and can</u> start testing now.

Correct: You have a clear description of <u>procedure and can</u> start testing now.

15.4.2 Introducing Elements

A comma is needed after an introductory word, phrase, or clause to set a brief pause before the main part of the sentence begins. This is especially important to pace control for the sentences with long introducing elements. *See* Appendix and Table 12.1 for transitional words and phrases.

Example 15.5 Comma introducing elements

Transitional Word
- <u>However,</u> the SO_2 removal efficiencies met our expectations.

Transitional Phrase
- <u>For example,</u> the SO_2 removal efficiency reached 99.5% when the temperature was 40°C.

Introductory Long Phrase
- <u>During the final step of the pilot tests,</u> the cobalt-based solvent did not show the NOx removal efficiencies as expected.
- <u>Pilot tests finished,</u> and we realized that the cobalt-based solvent did not show the NOx removal efficiencies as expected.

Introductory Clause

- <u>Because the hospital did not have enough ventilators for all patient during the of COVID-19 pandemic,</u> health professionals had to make difficult decisions on who got the life-saving machines.

Note

A sentence that begins with an introductory element may change the emphasis of the sentence. (*See also* 12.8.1.) In addition, the comma is optional if it follows a short phrase (*see* Example 15.6).

Example 15.6 Comma omission from short phrases

<u>In pilot test the</u> cobalt-based solvent did not show the NOx removal efficiencies as expected. (*short phrase*)

15.4.3 Separating Parallel Elements

Commas are used to separate parallel elements (words, phrase, or clauses) in sentences. For example, *I like music, sports, travel, and writing*. However, comma is unneeded for a parallel structure with only two elements written in a sentence. When there are three or more elements in a parallel structure, including the last comma improves clarity, especially when the last comma comes right before the word *and* or *or*. (*See* Example 15.7.)

15.4.4 Enclosing Elements

You can use commas to enclose nonessential, appositive, or interrupting elements in sentences. (Appositive phrase means two phrases referring to the same thing.) If the nonessential elements enclosed by commas were removed from the sentence, the rest of the elements still construct a valid sentence, and the primary idea of the sentence survives the removal. (*See* Example 15.8.)

Example 15.7 Comma for separating elements in parallel series

- *Word Series*
  Vague*:*   Apple, Blackberry, <u>Huawei and Samsung</u> were the leading smartphone pro-
                      ducers in the world. (*With a comma following Huawei,* <u>*Huawei and Samsung*</u>
                      appears to be one company to some readers.)
  Clear: Apple, Blackberry, <u>Huawei, and Samsung</u> were the leading smartphone pro-
                      ducers in the world.
  Incorrect: I like <u>music, and sports</u>.
  Correct: I like <u>music and sports</u>.
  Vague: I like <u>music, travel and sports</u>.
  Clear: I like <u>music, travel, and sports</u>.
  Incorrect: You can choose <u>music, or sports</u>.
  Correct: You can choose <u>music or sports</u>.
  Vague: You can choose <u>music, travel or sports</u>.
  Clear: You can choose <u>music, travel, or sports</u>.

- **Phrase Series**
  To complete this course, students must learn the basics in the classroom, observe their applications on a training site, give presentations to the class, and submit final reports by email.

- **Clause Series**
  A modern university education is all about students: students are motivated in learning, staff
  members are keen to serving the students, and faculty members are passionate in teaching.

Example 15.8 Comma enclosing elements
Nonessential Element

- From the results in this paper<u>, which were presented in Section 4,</u> they drew the following
  conclusions. *[non-restrictive clause]*
- The research team<u>, working day and night,</u> finished the project on time. (*emphasis*)

Appositive Phrase

- The only book that Margaret Mitchell authored<u>, *Gone with the Wind*,</u> was immensely popular
  when first released.

Brief Pause by Interrupting Words or Phrases

- The revised manuscript<u>, therefore,</u> is accepted for publication.
- The revised manuscript<u>, however,</u> cannot be accepted for publication.

Comments
The words enclosed with commas (*therefore, however, for example*) slow down the pace of
presentations, and they interrupt the continuity of thought.

15.4.5 Using Commas with Numbers

Commas are commonly used with addresses, dates, and numbers in technical writing. *See also* 16.2.14 Formatting Numbers. For example, the Arabic number of *one million, two hundred thousand, and three hundred* is 1,200,300. However, many countries use periods (1.200.300) or single spaces (1200300) for the same number – they are rarely used in English documents.

Example 15.8.2. Comma with number, date, and address

The 1,234-page-long document was released to the public on April 1, 2020, in Waterloo, Ontario, Canada.

15.4.6 Replacing Verbs

Experienced writers may replace the verb in a sentence with a comma. You may have noticed that I occasionally did so in this book (e.g., *the passive voice is used frequently in some languages, but in others, not at all*). It is grammatically correct, but it chops the sentence into pieces. The omission of a verb may confuse non-native English readers. Thus, avoid replacing the verb in a sentence with a comma, especially in engineering academic writing.

Example 15.9 Comma replacing verb

Confusing: Some researchers <u>are</u> prolific in publication; others, not at all.
Clear: Some researchers <u>are</u> prolific in publication; <u>others are</u> not at all.

15.4.7 Comma Intrusion

Comma intrusion occurs when a comma appears where it is not supposed to. It is often resulted from transforming informal writing into formal writing. You may frequently pause in verbal commutations or informal writing to show that we are thinking, but you should avoid inserting comma in writing wherever you feel like a brief pause. Engineering academic writing must use commas without challenging grammatical rules of sentence construction. Additionally, comma intrusion interrupts the flow of ideas.

Example 15.10 Comma intrusion

Incorrect: Acidified rain, ozone, and particulate <u>matter, can all</u> have a devastating impact on the environment.
Correct: Acidified rain, ozone, and particulate <u>matter all can</u> have a major impact on the environment.

Especially, avoid a comma separating a subject from its verb or a verb from its object. A non-native English writer might think it emphasizes what follows. Instead, it becomes a grammatical error.

Example 15.11 Comma challenges (1)

Wrong: The director of the <u>laboratory, and</u> the administrative staff are responsible for the safety of the researchers working in the laboratory.

Right: The director of the laboratory and the administrative staff are responsible for the safety of the researchers working in the laboratory.

Wrong: The principal investigator designed the test <u>apparatus, and</u> gave a demonstration to the team members.

Right: The principal investigator designed the test <u>apparatus and</u> demonstrated it to the team members.

Other types of comma intrusion include inserting comma in a compound element consisting of only two elements and enclosing a coordinating conjunction (*and*, *but*, *or*) with commas.

Example 15.12 Comma challenges (2)

Wrong: The principal investigator designed the test <u>apparatus, and, gave</u> a demonstration to the team members.

Right: The principal investigator designed the test apparatus <u>and</u> demonstrated it to the team members. (*It is clear who gave a demonstration to the team members.*)

Wrong: Avoid comma between an object and its verb<u>, or,</u> a verb and its object.

Right: Avoid comma between an object and its verb <u>or</u> a verb and its object.

In addition, avoid joining two independent clauses with one single comma. Many non-native English writers run two independent clauses together using a comma instead of a period. This results in a *run-on sentence* or a *comma splice*.

On the other hand, do not omit commas when they are needed. For example, a comma should immediately follow a conjunctive adverb or a conjunctive phrase that joins two independent clauses. (*See* Table 12.1. Conjunctions).

Example 15.13 Comma challenges (3)

Incorrect: We turned on the <u>power, the</u> heater started heating the reactor.

Correct: We turned on the <u>power. The</u> heater started heating the reactor.

Correct: <u>After</u> we turned on the power<u>, the</u> heater started heating the reactor.

Correct: We turned on the power<u>, and</u> the heater started heating the reactor.

Correct: We turned on the power<u>; the</u> heater started heating the reactor.

Example 15.14 Comma challenges (4)

Wrong: The author presents a timely work; <u>nevertheless it</u> does not match the scope of the journal.

Right: The author presents a timely work<u>; nevertheless, it</u> does not match the scope of the journal.

15.5   Dash

A dash is a short line longer than a hyphen or two consecutive hyphens. It indicates linkage, separation, emphasis, informality, or abruptness. However, it is only occasionally used in engineering academic writing. You can write an engineering document without dash marks.

15.6   Ellipsis

The ellipsis mark (…) indicates the omission of content. It is rarely used in scholarly technical writing because omission may reduce clarity. In this book, however, I use several ellipsis marks to show the omission of content that I quote, which are not essential to my writing in this book. You should avoid quotation in engineering academic publications aimed at research.

15.7   Hyphen

A hyphen (−) is used primarily in creating compound words; it can also act as a linkage or an emphasis to improve the clarity of writing. You can find numerous hyphenated words in this book, but the hyphen has more functions as follows.

15.7.1 Hyphenated Words

Multiple words joined by one or more hyphens form a compound word. For example, *state-of-the-art*, *three-quarters*, and *forty-eight*. Standard hyphenated words are collected in dictionaries. You can also create new hyphenated words as needed when you write.

15.7.2 Multi-word Modifiers

You can create a two- or three-word adjective modifier having one single meaning. Multi-word modifiers precede nouns.

Example 15.15 Hyphenated words as modifiers

Springer Verlag is a <u>well-thought-of</u> publisher for engineering books and periodicals.

15.7.3 Series

You can use hyphens in a series of unit modifiers that precede the same noun. Then, your writing flows smoothly with brevity.

> **Example 15.16 Hyphenated words in series**
>
> You can create <u>two- or three-word</u> "adjective" modifiers, which have single meanings.

15.7.4 Prefixes

A hyphen can be used between a prefix and a proper noun or a word acting as a noun (e.g., co-op and co-chairs). It is also used when the prefix ends in the same vowel that the root words start with (e.g., *semi-industrious, and pre-engineered*).

> **Example 15.17 Hyphenated words as prefixes**
>
> Power plants reduce air emissions by <u>pre-, in-, and post-combustion</u> control technologies.

15.7.5 Hyphens for Clarity

You can use hyphen to improve clarity and avoid ambiguity. The hyphens inserted into different words may lead to different modifiers.

> **Example 15.18 Hyphenated words for clarity**
>
> **Ambiguous**
>
> - The City of Waterloo is investigating the feasibility of a <u>biochemical waste management</u> facility.
>
> **Clear**
>
> - The City of Waterloo is investigating the feasibility of a <u>biochemical waste-management</u> facility.
> - The City of Waterloo is investigating the feasibility of a <u>biochemical-waste management</u> facility.

15.8   Parentheses

Words, phrases, or sentences enclosed by parentheses normally clarify or define the preceding texts. The information contained in the parentheses is nonessential to the sentence, but it may be helpful to some readers. However, you can remove the materials in parentheses from the sentence, and the key idea of the sentence should survive the removal.

> **Example 15.19 Parentheses in sentences**
>
> Pre-combustion (prior to combustion) air cleaning is more cost-effective than post-combustion (after combustion) air cleaning.

The following is typical misuse of parentheses in formal writing.

1. Using parentheses with numeral for a number described with the words. This redundancy interrupts the flow of reading.

> **Example 15.20 Parenthetical error (1)**
>
> **Redundant:**   The abstract of a master's thesis should be less than two (2) pages.
> **Concise**:   The abstract of a master's thesis should be less than two pages.

2. Using parentheses within parentheses. Combining parentheses with bracket helps reduce ambiguity.

> **Example 15.21 Parenthetical error (2)**
>
> **Incorrect**:   (A manuscript submitted to *Nature* (January 2013))
> **Incorrect**:   [A manuscript submitted to *Nature* (January 2013)]
> **Correct**:   (A manuscript submitted to *Nature* [January 2013])

15.9   Period

Again, period and comma are the two most frequently used punctuation marks in engineering publications. A period ends a full sentence. Periods are also used in decimal points with numbers (e.g., 3.14; 9.81) and with lowercase abbreviations (i.e., etc., *a.m.*). Avoid ending a sentence with more than one period mark. This is often confusing to non-native English writers, especially when the sentence ends with an abbreviation that ends with a period. In addition, avoid a period mark where it does not belong; otherwise, it creates sentence fragments.

Example 15.22 Single period at the end of sentence
- The group meeting will start at 8:01 <u>a.m.</u>
- Scholarly publications include scientific journal articles, nonfiction books, theses, <u>etc.</u>

15.10   Question Mark

The question mark is rarely used in engineering publications, which are mostly narrative. However, it may appear in direct quotations, or it is occasionally used to provoke discussion.

15.11   Quotation Marks

15.11.1   Enclosing Elements

The quotation marks may indicate a meaning of *so-called* when words, phrases, and clauses are enclosed. However, enclosed sentences and paragraphs indicate direct quotations. (*See* 2.2.)

Example 15.23. Quotation marks meaning *so-called*.
Many people believe that hydrogen is a <u>"zero-emission"</u> fuel.

15.11.2   Titles and Quotation Marks

The actual titles of the documents are not enclosed by quotation marks. In the body text, however, quotation marks are needed to enclose titles of most publications, except for titles of books and periodicals. The titles of books and periodicals normally appear in *italic* font.

Example 15.24 Quotation marks enclosing title

- Einstein's doctoral thesis, "A New Determination of Molecular Dimensions", published in 1905 was only 12-page long.
- "Basics of Gas Combustion" is the third chapter of the book titled *Air Pollution and Greenhous Gases*.

15.11.3   Punctuation in Quotation Marks

Materials directly quoted remain in the quotation marks, except for the following two fixed rules:

- Colons (:) and semicolons (;) are *outside* closing quotation marks.
- Commas (,) and periods (.) are *inside* closing quotation marks.

15.12 Semicolon

Semicolons used in long sentences can improve clarity and control the pace of presentation. Use a semicolon where a brief pause is needed; the pausing effect of the semicolon is between the effects of a comma and a period. A semicolon separates two independent clauses when they are of equal importance.

You also can replace the conjunctive word *and* with a semicolon to improve clarity. Example 15.25 compares the draft with the revised, where semicolon is used to improve clarity.

In complex sentences, semicolons may precede transitional elements, conjunctive adverbs, or coordinating conjunctions.

Example 15.25 Use semicolon to improve clarity

Complicated

In a typical operation, an appropriate amount of heterogeneous solid catalyst (with particle sizes between 420 and 840 μm) is preloaded and supported inside the tubular reactor as a packed bed between two quartz wool plugs at the upper 2/3 length of the reactor (20 cm, reaction zone) <u>and</u> the lower 1/3 length of the reactor (10 cm, preheating zone) remains empty to preheat the flowing biphasic media to the predetermined reaction temperature before entering the reaction zone.

Revised with Improved Clarity

In a typical operation, an appropriate amount of heterogeneous solid catalyst (with particle sizes between 420 and 840 μm) is preloaded and supported inside the tubular reactor as a packed bed between two quartz wool plugs at the upper 2/3 length of the reactor (20 cm, reaction zone<u>); the</u> lower 1/3 length of the reactor (10 cm, preheating zone) remains empty to preheat the flowing biphasic media to the predetermined reaction temperature before entering the reaction zone (Souzanchi et al. 2021).

Example 15.26 Semicolons separating elements
Transitional Words or Phrases

The basics of heat transfer are important to the design of key components of <u>powerplants; for example</u>, boilers, economizers, and waste-heat recovery units all have heat exchangers.

Conjunctive Adverbs

The computational models are still running<u>; therefore,</u> we do not know whether they agree with the experiments.
COVID-19 has caused anxiety in Canada<u>; meanwhile,</u> the government reactions did not help much at the very beginning.

Coordinating Conjunction

The editorial office may invite qualified reviewers from academics, national labs, and large corporations<u>; but,</u> as I said, not all of them will accept the invitations.

The preceding example shows that it is acceptable to use a semicolon before the conjunction instead of a comma if at least one of the independent clauses of a compound sentence contains commas.

Similarly, a semicolon may also separate elements in a series when one or more of the items contain commas. It avoids the confusion and achieve smooth flow of thoughts. Do not follow a semicolon with *and* or *or*, although you can do so for comma.

Example 15.27 Semicolons separating items in a series

Semicolon: The co-authors of this article are <u>Bin Zhao, Tsinghua</u> University; Chun Chen, Chinese University of Hong <u>Kong; and Chao</u> Tan, University of Waterloo.
Comma: The co-authors of this article are <u>Bin Zhao, Chun Chen, and Chao Tan</u>.

After revising the entire document three times, you should be satisfied with the language and the technical contents in your writing. Meanwhile, you may have formatted most part of the document based on your earlier experience while you were revising your document. Note that you formatted the document with distractions because your attention was focused on the language and technical contents. You now need to revise your document one more time, concentrating only on format.

15.13 Practice Problems

Question 1   Improve clarity of writing by precise punctuation; improve clarity and conciseness of writing by other revisions as you see necessary.

1. However, smaller fines succeed in worming into the downstream catalytic trickle bed. As a matter of fact, even low concentration of fine solids, in the range of several parts per million in the hydrotreater feed, may cause, over months of operation, substantial deposition throughout the catalyst bed. (*Edouard* et al., *2006. Chemical Eng Sci 61: 3875–3884*)
2. Atmospheric aerosols, also known as the airborne particulate matters (PM), have a significant impact on the atmospheric environment, mainly reflected in the aspects of: visibility degradation, public health, climate forcing, and hydrological effect.
3. When doped with 270 nm PDA nanospheres, the "necklace" nanostructure yielded best capture efficiency, 50–70%, for four different kinds of cancer cells.
4. Jin et al. [38] summarized the label-free methods, which rely only on the biophysical properties of CTCs and no chemical labeling process is involved. Wang et al. [39] also summarized the nanostructure-based platforms for CTCs enrichment. This work [it is unclear what this represents, Jin et al. or Wang et al.?] covered nanostructure fabrication and its application in CTC capturing, and explained and the capture mechanism. Dong et al. [40], as another example, introduced microfluidics-based platforms, from ligand-free to ligand-based microchannels. The authors pointed out the promising future of hybrid strategies in the end.
5. It was shown that more than 95% of cancer cells were adhered to the nanorough part when these cancer cell suspensions were incubated with a patterned substrate (half is smooth while the other half is nanoroughened).
6. The separation mechanism lies in the different hydrodynamic forces between CTCs and blood cells (cell size dependent) when they flow through these microchannels.

7. Lithium-ion battery is an important energy storage and transmission technology in a modern society. Due to the features of high energy density, long cycle life and low cost, LIB has been widely used from portable electronics, electrical vehicles to large-scale power sources (Li and Tan 2020).

8. Figure 3(a) compares the effects of the separator thickness and porosity on the battery energy density (the effect of separator thickness is shown in the red line, while the separator porosity is shown in the black line). The battery energy density dropped quickly from 148.8 to 110.6 Wh/kg with increased separator thickness from 5 to 100 μm. While, the impact of the separator porosity on battery energy density is not significant in this study. The discharge profiles for the batteries with different separators are also presented in Fig. 3(b) and (c). When increasing the separator thickness from 5 to 100 μm, the working voltage of the discharge plateau was decreased by about 0.02 V and the discharge capacity was only 56.3% of its original. Changing the separator porosity from 35% to 95% does not have significant impact on the battery discharge profile, with only slight decrease of the working voltage was observed (Li and Tan 2020).

9. The samples of 0.5MHNO, 0.97BWO, 2.99BWO, 4.99BWO only showed characteristic peaks of Bi_2WO_6, and the intensity of the strongest (131) peak at 28.299° changed significantly from about 1572 to about 10,631 as the acidity decreased.

10. The UV-Vis DRS was used to determine the photo-response range of nanomaterials synthesized under different acid and alkali conditions. It can be seen that all samples demonstrated high photo-absorption at around 380 nm, and their absorption bands were approximately 450 nm. Therefore the synthesized samples are all responsive to visible light regardless of the pH value of the precursor. Furthermore, the optical band gap of the as-prepared photocatalysts were calculated using Kubeka-Munk method, and the results are summarized in Table 5 (Liu et al. 2020).

11. As the acidity decreases, regular large nanosheets disappear, and are replaced by irregular flocs consisting of small nanosheets for precursor prepared in 0.5 M HNO_3 solution.

12. Moreover, this is still a relatively new method with the first systematic study reported in 2013 and research in this area still focuses on characterization of nanoroughened surfaces (e.g., stiffness, morphology, etc.) and interactions between tumor cells and nanostructure (e.g., guiding effect, focal adhesion, etc.).

13. Micro−/nano-scale magnetic beads grafted (mostly by streptavidin-biotin links) with ligands, are used to recognize and bind with CTCs.

Finalization

16

16.1   Repeated Revisions

There are many steps to finalize your document, including formatting and proofreading. In addition, you need to secure permissions to copyrighted materials before publishing your document on any platform.

Make sure that you have exhaustively revised your manuscript many times, reaching your limit of capability. Revision is time-consuming, and you may need to repeat it many times for the entire document. Your efforts can gradually refine your manuscript to the level of your satisfaction.

First, you begin revision as soon as you finish your first draft. The rough first draft of your document has numerous writing errors, and this is normal for most writers. In your first revision, you concentrate your attention on three *C's*, *clarity*, *conciseness*, and *coherence*, when you revise your document. Read through the document and ensure smooth transitions between paragraphs without interrupting your flow of thought. Pay attention to the logics of your writing, and make sure that your evidence and argument really support your statements and conclusions. Your writing should deliver exactly what you mean.

Then, you revise the document multiple times by considering grammar, spelling, structure, punctuation, and so forth. Grammatical correctness is the basics of academic writing. Precise wording is also important to reducing ambiguity (*see* Words and Phrases). Meanwhile, carefully check the punctuation.

Before formatting, you repeatedly read through the document to identify and remove the redundant materials from the document (*see* Example 16.1). Again, academic writing is characterized with clarity and conciseness, especially after three rounds of revisions.

Example 16.1. Removal of Redundant Materials

Redundant

- The *application* of a heterogeneous photocatalyst has been successful in <u>two *applications.*</u>

Correct

- ~~The application of a h~~Heterogeneous photocatalyst has been successful in two applications.
- The application of a heterogeneous photocatalyst has been successful in *<u>two industries</u>*.

All co-authors, if any, should contribute to revising the document. They may participate at different stages or with different depths, but the co-authors should do so prior to formatting and proofreading.

16.2 Formatting

If outlining a document is like framing a house, then formatting the document is like furnishing the house. After you have put so much effort to drafting and revising your document, it is time to bring its quality and readability to the highest standard possible.

Most publishers provide the authors with guidelines on manuscript format. The guidelines include step-by-step instructions on formatting titles, headings, abbreviations, symbols, captions, visuals, equations, body text, and so on. Follow their instructions closely if you have identified the publisher; otherwise, you may follow the general guidelines described in this chapter.

16.2.1 Formatting Title

Spell out the words whenever it is possible and capitalize the following elements in the titles.

- First and last words
- Major words for a long document
- Long prepositions with *five* or more letters (*between*, *since*, *until*, and *after*)

Except fully capitalized titles, do not capitalize words for short documents such as journal articles, conference papers, and chapters in a thesis. Do not capitalize coordinating conjunctions or short prepositions either.

Example 16.2. Title Formatting of a Thesis and an Article

Incorrect: Recording clinical-grade <u>ECG</u> from the upper arm
Correct: Recording Clinical-grade Electrocardiogram from the Upper Arm (a *thesis title*)
Incorrect: A review on the catalytic combustion of soot in <u>Diesel</u> particulate filters for automotive applications: <u>From</u> powder catalysts to structured reactors (Fino et al. 2016)
Correct: A review on the catalytic combustion of soot in <u>diesel</u> particulate filters for automotive applications: <u>from</u> powder catalysts to structured reactors

---

16.2.2 Formatting Table of Contents

Most Tables of Contents (TOCs) can be automatically created by modern word processing software. The software ensures that major *headings* and *subheadings* are included in the TOCs with default formats. Typical word processing software provides many TOC styles. Each style is designed for a specific group of readers. You can choose the right style to start with and refine it as needed. You may consider formal style for a formal document, like the one used in this book. It looks conservative, but professional. However, you often need to refine the TOC created by the software. For example, font, alignment, and line spacing are typical points of attention.

16.2.3 Justification of Margins

Justification of margins is a personal choice. Some like left-justified margins, and others prefer fully justified margins. Fully justified text appears formal, polished, and important. On the other hand, left-justified margins avoid uneven space between words. Furthermore, left justification is easy to process for software like *Microsoft Office*. Nonetheless, you should avoid center alignment of paragraphs in the body, although visuals and some headings may be center aligned (*see* Example 16.3).

Stretching or condensing words to fit them into one line distorts the letters, numbers, punctuation marks, and so on. Instead, slightly adjust the text by scaling up or down proportionally to fit the words into one line.

Example 16.3. Justification of Margins

Left Alignment (White Spaces on the Right)
Refrain from stretching or condensing words to fit them into one line. It distorts the letters, numbers, punctuation marks, and so on. Instead, slightly adjust the text by scaling up or down proportionally to fit the words into one line.

Full Alignment (Large Space Between Words)
Refrain from stretching or condensing words to fit them into one line. It distorts the letters, numbers, punctuation marks, and so on. Instead, slightly adjust the text by scaling up or down proportionally to fit the words into one line.

Center Alignment (Awkward)
Refrain from stretching or condensing words to fit them into one line. It distorts the letters, numbers, punctuation marks, and so on. Instead, slightly adjust the text by scaling up or down proportionally to fit the words into one line.

16.2.4 Line Spacing

Line spacing is important to the legibility and comfort of reading. The space between lines should be just right; either too tight or too loose distorts the page layout.

Applying the same line spacing to the entire document ensures consistency in format. The line spacing should range from 1 to 1.5 lines. You may use double-line spacing for drafting and revision, but not the final document. However, minor adjustments are acceptable to fit contents into one page.

Last, read through the document and check whether any heading or subheading is located at the bottom of a page, or any visual (including caption title) is separated onto two pages. You can fix these layout errors by carefully adjusting the line spacing.

Example 16.4. Line Spacing

Normal Spacing (1.2 Lines)
Read through the document and check whether any heading or subheading is located at the bottom of a page, or any visual (including caption title) is separated onto two pages. You can fix these layout errors by carefully adjusting the line spacing.

Too Tight (1 Line)
Read through the document and check whether any heading or subheading is located at the bottom of a page, or any visual (including caption title) is separated onto two pages. You can fix these layout errors by carefully adjusting the line spacing.

Too Loose (2 Lines)
Read through the document and check whether any heading or subheading is located at the bottom of a page, or any visual (including caption title) is separated onto two pages. You can fix these layout errors by carefully adjusting the line spacing.

16.2.5 Formatting Headings

Headings in the final document should be concise but descriptive for the entire section. Avoid too many or too few words in the headings because too many words cause clutter and too few words reduce clarity.

Format the headings with consistency and agreement. Certainly, you can be creative in heading styles, but they should be consistent throughout the document. Ensure that the font size, typefaces,

Example 16.5. Formatted Heading Example

1. **INTROUCTION**
   1.1 Literature Review
       1.1.1 *Motivation*
       1.1.2 *Experimental works*
       1.1.3 *Numerical works*
   1.2 Knowledge Gaps
   1.3 Objectives
   1.4 Thesis Structure
2. **METHODOLOGY**
   2.1 Overall Experimental Setup
   2.2 Test Apparatus
   2.3 Instrumentation and Data Collection
   2.4 Data Analyses and Expected Results
3. **RESULTS AND DISCUSION**
4. **CONCLUSIONS**

and capitalizations at the same level agree with each other. However, the lower-level headings normally differ from upper-level headings by size or style (such as bold face, capitalization, and italics).

Decimal numbering systems as seen in Example 16.5 appear in most engineering writing, including journal articles, books, reports, and theses. Nonetheless, some publications also use combinations of numbers and letters.

You can create a typographic contrast between headings and the text to improve clarity. You can achieve effective contrast with large font and bold face. However, all headings decrease font size in accordance with their levels. Thus, the font size of the lowest-level heading equals that of the body text.

Keeping the preceding principles in mind, you check your final formats following this checklist.

- Identical typeface for the same level of heading
- More capital letters in a higher level of heading
- More numerical numbers in a lower level of heading
- New typeface or smaller font size at a lower level of heading
- Bolded face at the higher level of headings
- All headings flush left margin

Furthermore, you should _avoid_ the following styles while format headings:

- **Lonely headings** (at the bottom of pages). You can adjust line spacing to move the heading onto the next page.
- **Indented headings**. Decimal headings flush with the left margins in the final document, although they are usually indented in the TOC (Table of Contents).
- **Indenting first paragraph** immediately following the heading. However, indent other paragraphs under the same heading or subheading, unless you separate two consecutive paragraphs by a blank line.
- **Suppressing headings**. In contrast, use double space for the headings.

16.2.6 Formatting Headers and Footers

Use concise words, phrases, and clauses in headers and footers. For both, you use concise elements with font sizes that are smaller than the main text to avoid visual clutter.

In addition, you need to pay attention to the format of page numbers in footers. Page numbers in the front matter should be consecutively numbered with the Roman numerals using lowercase letters (i, ii, iii…). The page numbers of the main body and backmatter are numbered using Arabic numbers (1, 2, 3).

16.2.7 Typeface and Font Size

A typeface is the overall design of the letters used in your writing. It can be **bold**, _italic_, underline, and so on. Your font changes with the typeface. There are many typefaces available, but you should use professional typefaces in technical writing that most readers are familiar and comfortable with. Commonly used typefaces in formal writing include Arial, Calibri, and Times Roman. Meanwhile, you should avoid strange typefaces that are illegible or childish.

Example 16.6. Different Typefaces

Avoid: typefaces such as *COMIC*, *Bradley*, and *brush*

Use: typefaces such as Arial, Calibri, Garamond, and Times New Roman

Formal publication requires consistent typefaces throughout a single document. However, you can use variants to achieve emphasis and contrast. For instance, the titles, headings, and subheadings normally use distinctive typefaces.

Size wise, most writers and publishers use 11 or 12 points as the default font size for the main text. However, the sizes of headings are normally larger, and those headers and footers are smaller than the main text size.

16.2.8 Capitalization

Numerous grammatical rules are related to appropriate capitalization. To start, you can refer to Table 16.1 for your writing. More can be found in handbooks for writing (e.g., Alred et al. 2018). You also need to avoid capitalizing the following words:

- The first word enclosed in dashes, brackets, or parentheses
- The first word after a colon (:) or semicolon (;)
- Autumn, spring, summer, and winter when they are used as seasons of the year
- Earth, sun, moon, and the like except when they mean astronomical bodies
- East, north, south, and west when they are used for directions instead of places

Example 16.7. Capitalization Challenge

Incorrect:	We must repeat the <u>tests (Data</u> just collected are lost).
Correct:	We must repeat the <u>tests (data</u> just collected are lost).
Incorrect:	We must repeat the <u>tests; Data</u> just collected are lost.
Correct:	We must repeat the <u>tests; data</u> just collected are lost.
Incorrect:	We must repeat the <u>tests: Data</u> just collected are lost.
Correct:	We must repeat the <u>tests: data</u> just collected are lost.
Incorrect:	Heading <u>East</u>, you will see a tall silver building.
Correct:	Heading <u>east</u>, you will see a tall silver building.

16.2.9 Capitonyms

Capitonyms may be challenging to some non-native English writers. Pay attention to capitonyms when you revise your manuscript. Capitonyms are dual meaning words, which change their meanings if the words are capitalized (Table 16.2).

Table 16.1 Capitalization

Capitalization	Example
Addresses and places	Beijing, China; Ontario, Canada
Associations	Canadian Society of Mechanical Engineering
Beginning of sentences	–
Corporations	Blackberry, BP, Microsoft Office
Cross-references	Chapter 5, Eq. 10, Section 2.3, Fig. 1, Table 6
Days and months	Monday, December
Ethnic groups	Chinese, European
Historical events	World War II
Institutions	Tsinghua University, University of Waterloo
Internal units of corporations, etc.	Faculty of Engineering US Department of Energy
Letters for shapes	*I*-beam, *T*-shape, *U*-turn
Names of people	John Lennon, Tom Smith
Nationalities	American, Arabic, Canadian, Chinese, German
Religions	Christianity, Jewish, Muslim

Table 16.2 An incomplete list of capitonyms

Capitalized		Lowercase	
Word	Meaning	Word	Meaning
China	the country	china	porcelain
Turkey	the country	turkey	the bird
March	the month	march	to walk (verb)
Titanic	the ship	titanic	gigantic (adjective)
Bill	name of a person	bill	amount to be paid
Lent	period in the Christina calendar	lent	past tense of *lend*
Reading	a town in England	reading	progressive form of *read*
Polish	relating to Poland	polish	smoothing by rubbing

16.2.10 *Italicizing* Words with Non-English Roots

You should italicize the words and the phrases with non-English roots, including their abbreviations. Most of these words have Greek or Latin roots. Look up in dictionaries when you are in doubt, but the following words and phrases frequently appear in engineering writing (Table 16.3).

16.2.11 *Italics* to Specify Items

An italicized element (words, phrases, symbols, and so forth) specifies that item. You see italicized sentence elements throughout this book, for example:

Example 16.8. Italicizing Specified Elements in Sentence

The word *italicize* means to *use italics*; it is not the language or the country because the initial *i* is in lowercase.

Table 16.3 Words and phrases with non-English roots

Root	Abbreviation	Meaning	Use in writing
exempli gratia	*e.g.*	for example	Usually enclosed in parenthesis. Examples can be found throughout this book
et alii	*et al.*	and others	Normally used to end an incomplete list of names in citations and references.
et cetera	*etc.*	and so forth	Used in incomplete lists in sentences. Examples can be found throughout this book
id est	*i.e.*	that is	*See also* Example 4.52
versus	*vs.*	against	Used to show relationship of one against the other (*e.g.*, Fig. 5e shows the change of density *vs.* pressure.)
vice versa	–	the other way around	Normally used to end a sentence following ", and" (, and *vice versa*)

16.2.12 Formatting Date

The format of date varies with country. In the USA, dates are usually written in the format of *month day, year*. In Canada, however, dates are written in the format of *day month year* (nothing in between). On the contrary, dates follow the format of *year month day* (nothing in between) in Chinese literature.

Regardless of the format, a date normally follows the preposition *on*, and the preposition *in* precedes a month only. When year is omitted, you should use the cardinal number (*January 4*) instead of the ordinal number (*January 4th*). Nonetheless, avoid commas when day is not seen in the date.

Example 16.9. Format of Date (1)

USA: The data were collected on January 4, 2020.
USA: The last day of this semester is August 30, 2020.
Canada: The data were collected on 4 January 2020.
Canada: The last day of this semester is 30 August 2020.

Example 16.10. Format of Date (2)

We submitted the manuscript on <u>January 4</u>. The target date of resubmission is <u>February 30</u>.

Example 16.11. Format of Date (3)

The data were collected <u>in January 2020</u>.
The target date of submission is in <u>August 2020</u>.

Finally, you should avoid numerical-only form for dates in formal writing. Otherwise, it creates confusion or ambiguousness, especially to non-native English speakers. For examples, *10/2/20* may be interpreted as *October 2, 2020*; *February 10, 2020*; *2 October 2020*; or *10 February 2020*, depending on the country. Therefore, spelling out the <u>month</u> ensures clarity of writing.

Example 16.12. Format of Date (4)

Incorrect:	We submitted the revised manuscript on <u>10/2/20</u>.
Clear:	We submitted the revised manuscript on <u>October 2, 2020</u>.
Clear:	We submitted the revised manuscript on <u>2 October 2020</u>.
Clear:	We submitted the revised manuscript on <u>10 February 2020</u>.
Clear:	We submitted the revised manuscript on <u>February 10, 2020</u>.

16.2.13 Formatting Time

Time specifies the exact time of action. It may be an important factor for research such as atmospheric air dispersion and solar energy. Colons are used to separate hours from minutes in time followed by abbreviations *a.m.* or *p.m.* For example, 8:30 a.m. and 12:30 p.m. You need to spell out the words of time when *a.m.* or *p.m.* is absent. For example, 8:30 in the morning, 12:30 in the afternoon, and 7 o'clock in the evening.

16.2.14 Formatting Numbers

Many non-native English writers use numerals (1, 2, 100) for numbers in formal writing, which is a mistake. In formal writing, you need to write words for *integral* numbers ranging from zero through ten (*one*, *two*,...*ten*), and use numerals for numbers larger than ten. Meanwhile, you use commas to separate the elements of Arabic numbers greater than 999; the format is like 1,234,567,890. However, equations always use numerals for numbers, with decimals (*3.14*; *1.618*), and with units.

Example 16.13. Format of Number (1)

Informal:	I have published <u>3</u> books and <u>80</u> articles, and my goal is to publish <u>10</u> books and <u>150</u> articles.
Formal:	I have published <u>three</u> books and <u>80</u> articles, and my goal is to publish <u>ten</u> books and <u>150</u> articles.

For sentences beginning with numbers, use words for the numbers. You also can reconstruct the sentences to avoid starting with numbers. When one number precedes another in the same phrase, spelling out either one enhances clarity.

Example 16.14. Formatting Numbers

Incorrect:	<u>400</u> attendees participated in the annual event.
Correct:	<u>Four hundred</u> attendees participated in the event.
Rewrite:	<u>This year, 400</u> attendees participated in the annual event.
Unclear:	Our samples include <u>12 6-meter</u> long steel bars.
Revised:	Our samples include twelve 6-meter long steel bars.
Revised:	Our samples include <u>12 six-meter</u>-long steel bars.

16.2.15 Formatting Units and Symbols

Use the International System (SI) of Units in your writing unless you are clearly instructed otherwise. SI units are widely used in scientific and engineering publications, especially those aimed at international readers. Meanwhile, most engineering publications use SI units. The following grammatical rules of unit are generally applicable internationally.

- Leave one space between the measurement value and its unit.
- Use a small dot (·) or period (.) between two units to indicate multiplication operation.
- Use a backslash (/) between two units to indicate division operation.
- Avoid following an abbreviated unit with a period mark.
- Avoid adding letter "s" to an abbreviated unit for plural.
- Capitalize abbreviated units named after people (*see* Table 16.4; Wikipedia.com); otherwise, the abbreviated units are written in lowercase.

16.2.16 Formatting Lists

Lists are used more frequently in longer documents. There are lists in almost all technical books, dissertations, formal reports, and the like. However, you should be prudent with list if page limit is imposed on your writing.

When you format a list, carefully check and make sure that the entire list maintains parallel structure. All listed items normally begin with capital letters; listed sentences should also end with period marks.

Check that the items in lists follow the order you intended. A combination of numbers and letters allows you to list with subdivisions. For bullet points, bullets are used for items without clear rank or sequence. However, you may still want to list the items in a logical order that supports your purpose of writing.

Last, you should avoid ending items with commas or semicolons avoid using the word *and* or *or* to join consecutive items.

Table 16.4 Symbols and units named after pioneer researchers

Symbol	Unit	Named after	Used for
A	ampere	André-Marie Ampère	Electric current
C	coulomb	Charles-Augustin de Coulomb	Electric charge
°C	degree Celsius	Anders Celsius	Temperature
Hz	hertz	Heinrich Rudolf Hertz	Frequency
J	joule	James Prescott Joule	Energy
K	kelvin	William Thomson Kelvin	Temperature
N	newton	Isaac Newton	Force
Ω	ohm	Georg Simon Ohm	Electric resistance
Pa	pascal	Blaise Pascal	Pressure
V	volt	Alessandro G. A. Anastasio Volta	Electric potential (voltage)
W	watt	James Watt	Power

16.2.17 Formatting Visuals

Chapter 14 introduces basic visuals. However, you might have changed their formats and locations while revising the manuscript. Now, you need to relocate certain visuals and refine their formats before proofreading.

Double-Check Placement of Visuals

First, check and ensure that all visuals are placed with center alignment and that their captions are placed consistently: figure captions are under the figures and table captions are above the tables.

Then, make sure that all caption titles follow the same rules of grammar, spelling, and format as a *title*. However, they are not full sentences. So, you can omit nonessential articles.

Third, check and ensure that all visuals are integrated into the body text and they appear right after their first explanations. All visuals should be cross-referenced in the text, and vice versa. In addition, you may move paragraphs around while revising the entire document to further improve the coherence and unity. Now, check the relative locations of the explaining text and the visuals before finalizing your document for proofreading.

Occasionally, visuals are gathered in one place in a manuscript under review, but not for its final publication. For example, many international journals require authors to submit their manuscripts with figures and tables appended to the text-only documents, which contain only visual captions. However, that is only to facilitate peer review and editorial tasks. They may change in the final document.

Refining Visuals

Appealing visuals improve reading experience, which can positively impact your career advancement. Thus, format your visuals as if you were an artist. Admittedly, engineers do not pretend to be artistic designers, but the following basics in visual design may further improve the overall quality of your visuals. With enough attention and continual practice, you should become more and more skilled in visual design and formatting.

As explained in Chap. 14, visuals are composed of basic elements such as colors, forms, lines, shapes, texts, and white space. Their forms may be one-, two-, or three-dimensional. By considering all these factors, your formatting improves the clarity of the main document.

Maintain a balance between consistence and variety of visuals. The variety of visuals improves clarity. Variations in color and font size in the same visual create contrast and hierarchy effects, emphasizing the important elements.

Do not clutter your visuals. Leaving enough spaces around lines, shapes, and texts increases readability. White space is important to the comfort of eyes too. You can increase comprehension of visuals by proper use of white spaces between visual elements and within individual elements. You are encouraged to read some books about the white space in visual design (e.g., Hagen and Golombisky 2013).

You can also find many nicely crafted visuals in the public domain. Examine them against the general guidelines introduced in this book and apply them to your writing.

Table 16.5 Cross-reference equations

Cross-reference	Meaning
Equation 6	The sixth equation
Eq. 6	Same as Equation 6
Equation 6–8	The eighth equation in Chapter 6
Eq. 6–8	Same as Equation 6–8
Equations 6–8	Equations 6, 7, and 8
Eqs. 6–8	Same as Equations 6–8

16.2.18 Formatting Equations

Formatting equations is a simple but tedious work. The following formatting guides apply to all engineering equations.

Number the equations consecutively throughout the final document. Most writers enclose the equation numbers with parentheses or brackets, which are on the right side of the equations and flush to the right margin. Leave at least three spaces between the equations and their numbers to avoid misinterpretation. In addition, indent all equations by *at least* half an inch from the left margin.

Thoroughly check the body text and ensure that equations are cross-referenced by equation numbers preceded with the word *Equation* or the abbreviation *Eq.*, which is capitalized (*see also* Table 16.5). Either one is acceptable, but it should be used consistently throughout the document. However, avoid *Eq.* and use *Equation* at the beginning of a paragraph. Furthermore, avoid inserting equations in the sentence unless the equations are part of *Conclusions*. Last, *Table of Equations* is unneeded in your document.

A long equation that cannot fit into one single line should be broken into multiple lines. *See* Example 16.15. The new lines should start with an equal sign or an operation sign (+ × ÷) that is *not* enclosed with parentheses or brackets. Meanwhile, align the left side of the first line and right side of the last line with other single-line equations on the same page. The lines in between can be centered or aligned by the right of equations. Nonetheless, the long equation is numbered as one equation.

Example 16.15. Alignment and Arrangement of Long Equations

The mass balance in the cubic volume can be established as

$$\Delta C\left(\Delta x \Delta y \Delta z\right)=\left(\Delta J_x + \Delta J_y + \Delta J_x\right)\Delta t \tag{6}$$

With Eqs. 1–3, Eq. 6 becomes

$$\frac{\Delta C}{\Delta t}=\frac{\left(C_x - C_{x+\Delta x}\right)u+\left[\left(-D_x\frac{\partial C}{\partial x}\right)_x - \left(-D_z\frac{\partial C}{\partial z}\right)_{x+\Delta x}\right]}{\Delta x}$$

$$+\frac{\left[\left(-D_x\frac{\partial C}{\partial y}\right)_y - \left(-D_y\frac{\partial C}{\partial y}\right)_{y+\Delta y}\right]}{\Delta y} \tag{7}$$

$$+\frac{\left[\left(-D_z\frac{\partial C}{\partial z}\right)_z - \left(-D_z\frac{\partial C}{\partial z}\right)_{z+\Delta z}\right]}{\Delta z}$$

Taking limit of an infinitesimal cube and time interval, Eq. 7 becomes

$$\frac{\partial C}{\partial t}=-u\frac{\partial C}{\partial x}+D_x\frac{\partial^2 C}{\partial x^2}+D_y\frac{\partial^2 C}{\partial y^2}+D_z\frac{\partial^2 C}{\partial z^2} \tag{8}$$

Italicized Symbols

Finally, thoroughly check the document to make sure all symbols are *italicized*, including the equations and the body text. *See* Example 11.9.

16.2.19 References and In-Text Citations

There are three principal formatting systems for in-text citations and references:

- American Psychological Association (APA) system (APA 2020)
- Institute of Electrical and Electronics Engineers (IEEE) system (IEEE 2020)
- Modern Language Association (MLA) system (MLA 2020)

APA and IEEE styles or their combination are normally used in engineering publications.

The standard formatting styles are meant to be guidelines, and they should not discourage your creativity. Be yourself within certain constraints; so is true for writing. Regardless of the styles you choose, they should be consistent throughout your document.

References

The page layout and the line spacing of *References* are the same as those of the document body. The heading *References* appears on a new page in a long document, but not necessarily the case for short documents such as articles in international journals.

Reference entries are listed without bullet or numerical numbers. Treat each entry of reference list as one single paragraph. Align the first lines of the entries to the left margin, but indent one-half inch from the left margin for the other lines to follow. *See* Example 16.16.

As seen in *References* of this book, the entries are listed in an alphabetical order by the first or sole author's last name. Multiple entries written by the same author(s) are listed in the order of publication time, from early to recent. List each author by last name, followed by initial(s).

Example 16.16. Layout of Reference List

Dolan R, Yin S, Tan Z. 2009. Design and evaluation of a subcritical hydrothermal gasification system. *Proceedings of 50th International Conference on Bioenergy*, December 1–3, 2009, Calgary, Canada.

Dolan R, Yin S, Tan Z. 2010. Effects of headspace fraction and aqueous alkalinity on subcritical hydrothermal gasification of cellulose. *International Journal of Hydrogen Energy* 35:6600–6610.

To start, you can follow these formatting requirements.

1. List all authors following the same pattern.
2. Separate authors with commas.
3. Stop at the sixth author (inclusive) for papers with more than six authors.
4. Omit the seventh author and those after. Instead, use et al. to represent the rest of the authors.
5. Capitalize only the first words for the titles of articles, conference papers, and the like.
6. Use lowercase without italics for titles of short documents (e.g., book chapters and journal articles).

7. Capitalize and italicize the titles of long documents (e.g., books, theses, and periodicals)
8. Separate the title and the subtitle in the same entry using a colon.
9. Avoid enclosing title or subtitle with quotation marks, parentheses, or the like.
10. Separate the list of authors, year, title, page range, and other elements using commas or periods.

Last, check and ensure that each entry of the reference in the list is cited in your text; likewise, each in-text citation corresponds to one and only one reference. However, bibliography does not require this matching because there are more bibliographic entries than their in-text citations.

In-Text Citations

In-text citations, which appear in the main text, give credit to others for their contributions to your writing. There are at least three forms of in-text citations: number in brackets, number as superscript, and (*author, year*) form. Most publishers use the *author-year* form for the upfront credit to the authors, but in-text citations in parentheses may interrupt the flow of writing and reading.

Pay attention to the difference between parenthetical citation and narrative citation. You can remove parenthetical citation from the sentence without affecting the ideas of the sentence, but not for narrative citation because the author's name becomes an element of the sentence. *See also* Example 16.17. Parenthetical citations and narrative citations for illustration of difference.

The exact form of in-text citation depends on the number of authors. Single-authored work takes the (*author, year*) format by default (e.g., Tan, 2005). For a work with two authors, cite both names joined by the word *and* (e.g., Tham and Roy, 2020). For a work with three or more authors, list only the last name of the first author followed by et al. (e.g., Tan et al., 2014).

When you enclose multiple citations using the same parenthesis, list them alphabetically and separate them with semicolons (e.g., Adam, 1971; Tham and Roy, 2020; Tan et al., 2014).

Example 16.17. Parenthetical Citations and Narrative Citations

Parenthetical Citations

-   Other researchers have reported the effects of headspace fraction and aqueous alkalinity on subcritical hydrothermal gasification of cellulose (Dolan et al. 2010) (*reduces interruption*).
-   Other researchers (Dolan et al. 2010) have reported the effects of headspace fraction and aqueous alkalinity on subcritical hydrothermal gasification of cellulose. (*Position in between interrupts the flow of reading.*)
-   Figure 5-5 shows a figure in the article recently published in the Journal of Power Sources (Li et al. 2019).

Narrative Citation

-   Dolan et al. (2010) reported the effects of headspace fraction and aqueous alkalinity on subcritical hydrothermal gasification of cellulose.
-   Figure 5-5 shows a figure in the article written by Li et al. 2019).

Recommended Styles for References and Citations

No single standard style pleases all readers. I recommend the following styles for long and short documents.

- **Articles in journals, magazines, *etc.***
  Author(s). Year. Title of article, *Title of Journal* Volume number (Issue number): page number range.
- **Peer-reviewed conference papers**
  Author(s). Year. Title of paper, *Proceedings of the Conference Name*, Conference time, Location.
- **Chapters in books and reports**
  Author(s). Year. Chapter title, in *Title of Long Document*, edition, Publisher Name, Address.
- **Works in collections**
  Author(s). Year. Chapter title, in *Title of Long Document* (editor: name), edition, Publisher Name, Address.
- **Books and large reports**
  Author(s). Year. *Title of Document*, edition, Publisher, Address.
- **Degree theses**
  Author(s). Year. *Title of Document*, <u>Doctoral</u> or <u>Master's</u> thesis, University Name, Country.

Table 16.6 summarizes the entries and the corresponding parenthetic in-text citations. They are primarily based on APA system, but with some modifications to improve conciseness and clarity.

16.3 Formatting Cross-References

A cross-reference directs readers to another place for related text in the same document. Section numbers and page numbers are optional cross-references, but they help the readers quickly find the related text.

Make sure all cross-references are formatted consistently throughout the document. It is optional to enclose cross-references in parentheses. Italicize the word(s) *see* or *see also*. When page number is included, precede the page number with a comma.

All cross-references should capitalize the first words, including *Equation/Eq./Eqs.*, *Figure/Fig./Figs.*, *Line*, *Page*, *Section*, and *Table*. In addition, avoid preceding cross-references with articles (*a*, *an*, *the*).

16.3.1 Formatting Appendices

Format the appendices following the same styles as in the body text. If you have only one appendix, label it as *Appendix*, otherwise *Appendices*. If you have more than one appendix, give each of them a heading (*Appendix A*, *Appendix B*, etc.). Do not use numerals in appendix headings. Instead, use letters *A*, *B*, *C*, and so on. You may generate a *List of Appendices* like the *List of Figures* for a long document. *See* Appendices in this book for example.

16.3.2 Formatting Index

Index is unnecessary to short documents, and it is optional to long documents. Many books or large reports have indices, but I found them rarely useful. In addition, it would be time-consuming to manually create the indices of your long documents. Nowadays, most word processing software allows you to create indices automatically. To start with, you may refer to *The Chicago Manual of Style* by the University of Chicago Press Editorial Staff (2017) for guidelines on index.

Table 16.6 Styles of reference entries and in-text citations

Type	Author	Example	Citation
Journal article	Single author	Tan Z. 2008. An analytical model for the fractional efficiency of a uniflow cyclone with a tangential inlet, *Powder Tech.* 183 (2): 147–151.	Tan, 2008
	Two authors	Givehchi R, Tan Z. 2015. Effects of capillary force on airborne nanoparticle filtration, *Journal of Aerosol Science* 83: 12–24.	Givehchi and Tan, 2015
	Three to six authors	Li J, Cheng K, Croiset E, Anderson WA, Li Q, Tan Z. 2017. Effects of SO_2 on CO_2 capture using chilled ammonia solvent, *International Journal of Greenhouse Gas Control* 63: 442–448.	Li et al., 2017
	Seven or more authors	Johnson MD, Fish VL, Doeleman SS, Marrone DP, Plambeck RL, Wardle JFC, et al. 2015. Resolved magnetic-field structure and variability near the event horizon of sagittarius A, *Science* 350: 1242.	Johnson et al., 2015
Paper	Same as above	Sun F, Sade B, Ghosh H, Tan Z, Sivoththaman S. 2019. Synthesizing reduced graphene oxide, *Proceedings of the 46th IEEE Photovoltaic Specialist Conference*, June 16–21, 2019, Chicago, IL, USA.	Sun et al., 2019
Work in collection	Same as above	Tan Z. 2013. Nanoaerosol, in *Encyclopedia of Microfluidics and Nanofluidics* (ed. Li D), 2nd ed., Springer Verlag, Singapore.	Tan, 2013
Collection	Same as above	Tan Z. 2019. *Micro/Nano Materials for Clean Energy and Environment*, special issue of *Materials*, MDPI, Switzerland.	Tan, 2019
Book	Same as above	Tan Z. 2014. Air Pollution and Greenhouse Gases, Springer Verlag, Singapore.	Tan, 2014
	Same as above	Ashrafizadeh SA, Tan Z. 2018. *Mass and Energy Balances*, Springer Verlag, New York, USA.	Ashrafizadeh and Tan, 2018
	Corporate author	Mathworks (2020). *Matlab Numerical Computing Tutorials.* Natick, MA, USA.	Mathworks, 2020
	Government author	DOE (Department of Energy). 1995. *Life-Cycle Costs for the Department of Energy Waste Management Programmatic Environmental Impact Statement* (Report INEL-95/0127 [Draft]). Department of Energy, Washington DC, USA.	DOE,1995
Thesis	Normally single author	Saprykina A. 2009. *Airborne Nano- particle Sizing by Aerodynamic Particle Focusing and Corona Charging.* MSc thesis, University of Calgary, Canada.	Saprykina, 2009
		Givehchi R. 2015. *Filtration of NaCl and WOx Nanoparticles Using Metal Wire Screen and Nanofibrous.* PhD dissertation, University of Waterloo, Canada.	Givehchi, 2015

16.4 Proofreading

Your manuscript is ready for proofreading now. Proofreading is the final step in academic writing for publication. Do not rush into proofreading; you should proofread only after you have finished all other writing tasks.

Proofreading focuses on surface errors in spelling, grammar, punctuation, and format. Meanwhile, you can identify and refine the imperfections in typefaces, spaces, alignments, and visuals.

Proofreading starts with spelling and grammar using word processing software or other tools. These tools, however, have limited database and rules. They can identify some spelling and grammatical errors, but not all. More importantly, they cannot check your organization of ideas. Thus, you must proofread the ideas, logics, tones, voices, emphasis, and so on – these are key factors that define *your* writing style.

Finally, make sure your final document looks professional, although impression is secondary to the clarity, coherence, and clarity of writing. Like it or not, the appearance of your document affects readers' interests. Some readers might judge your capability based on your writing. You may never know your readers, but your effort in proofreading pays off in the end.

Be patient and proofread slowly.

16.5   Copyright Clearance

Proofread and make sure that your document is free of plagiarism and that you secure all permissions to use copyrighted materials. Software such as *Turnitin* and *iThenticate* is a useful tool for plagiarism checking. While *Turnitin* is primarily for educational purpose, *iThenticate* is aimed at researchers to ensure the originality of scholarly publication. Ask your employer for the site license. Both *Turnitin* and *iThenticate* can generate similarity reports, which identify the possible plagiarism in your writing.

You must obtain written permission from the copyright holder to use copyrighted materials. You can request the permission by contacting individual copyright holder. You may submit your request online if the copyright holders are larger corporations like *Springer* and *Elsevier*. Either way, you need to secure permissions on a case-by-case basis.

16.6   Internal Clearance

You also need to secure the permission from your employer or supervisor to publish any work that may impact their business. Your employer may have signed a legal document to support a research grant, which you may not know. On campus, you should give a copy of the manuscript to your supervisor for review before sharing it with anyone external or submitting it anywhere for the consideration of publication.

You must secure approvals for publication from all co-authors too. It would be unethical to list a co-author, who is usually influential, without his or her approval. Regardless of the intention or excuses, such an action is not acceptable in academic community. You are expected to be professional in academic writing for engineering publication.

16.7   Practice Problems

Question 1: (Equations)   Revise these equations following the guide in this book.

1. $\text{WHSV}\left(hr^{-1}\right) = \dfrac{\text{Feed concentration} \times \text{Feeding Flow Rate}}{\text{Mass of Catalyst}}$

2.
$$\text{Glucose Conversion}(\%) = \frac{\text{Moles of glucose converted}}{\text{Initial moles of glucose}} \times 100\%$$

$$= \frac{\left[\left(C_{Glu}^{F} \times Q\right) - \left(C_{Glu}^{P} \times Q\right)\right] / M_{Glu}}{\left(C_{Glu}^{F} \times Q\right) / M_{Glu}} \times 100\%$$

$$= \frac{C_{Glu}^{F} - C_{Glu}^{P}}{C_{Glu}^{F}} \times 100\%$$

Question 2 Examine Equations 1–3 of Sample Paper #4 and suggest how to revise it following the guide in this book.

- **Sample Paper #4:** Ge H, Hu D, Li X, Tian Y, Chen Z, Zhu Y. 2015. Removal of low- concentration benzene in indoor air with plasma-MnO_2 catalysis system, *Journal of Electrostatics* 76: 216–221. Available online: https://doi.org/10.1016/j.elstat.2015.06.003

Question 3: (Cross-references) Revise the cross-references following the guidelines in this book.

1. Fig. 1 presents the classification of CTC enrichment methods.
2. Table. 1 summarizes the enrichment performance of featured single-modality methods.
3. The Fig. 5 shows that the computational model agrees with the experiment.
4. Results are presented in the Section 4 on page 10.

Question 4 Correct the mismatch of font sizes in this sample thesis.

- **Sample thesis:** Lei Cheng, 2010, *Lignin Degradation and Dilute Acid Pretreatment for Cellulosic Alcohols,* Master of Science Thesis, UNIVERSITY OF CINCINNATI, USA. Available online: https://etd.ohiolink.edu/apexprod/rws_etd/send_file/send?accession=ucin1282329715&dispositio n=inline

Part IV

Correspondence

Correspondence 17

17.1 Peer Review Process

Understanding the peer review process for academic publication helps you write a cover letter. The peer review process changes over time. In the past, an Associate Editor selected peer reviewers, who were normally the experts in the field. The peer review could take weeks to months, depending on the availability and responsiveness of the reviewers. The Associate Editor recommends to the Editor in Chief by considering the reviewers' comments. The decision can be one of these options:

- Accept as is
- Minor revisions
- Major revisions
- Rejection

Recently, the process has become more complex than before by adding a cover letter from the corresponding author and a pre-screening by the editorial office. These additional steps aim to effectively manage the review of many manuscripts (especially journal articles) received by the publishers and to reduce unnecessary peer review – peer reviewers are volunteers and busy people.

An initial assessment by the Editor in Chief, an Associate Editor, or a qualified editorial staff may result in one of these four decisions:

- Rejected without review
- Rejected and resubmission
- Sent to external review
- Recommended to another journal

Regardless of the decision, you will receive a timely notification to avoid delay to the publication of your work. The decision is normally based on the scope, quality, novelty, and impact of your work. Therefore, you need to address these in your cover letter.

© The Author(s), under exclusive license to Springer Nature Switzerland AG 2022
Z. Tan, *Academic Writing for Engineering Publications*,
https://doi.org/10.1007/978-3-030-99364-1_17

17.2   Cover Letter

17.2.1 Structure of Cover Letter

You normally need to write a cover letter when you submit a manuscript for consideration of publication. A cover letter includes essential information, and it is normally 1- to 2-page long. Example 17.1 is a sample cover letter.

Example 17.1 Sample Cover Letter

Zhongchao Tan, PhD, PEng
Professor, Dept. of Mechanical & Mechatronics Engineering
University of Waterloo
200 University Avenue West
Waterloo, Ontario, Canada N2L 3G1
Phone: (519) 888-4567
Email: tanz@uwaterloo.ca

April 1, 2020

Dear Dr. Smith,

On behalf of my co-authors, I am submitting the manuscript entitled "Effects of Separator on the Electrochemical and Thermal Performances of Lithium-ion Battery: A Numerical Study". We would like to have the manuscript considered for the publication in *Energy & Fuels*. All co-authors have contributed to the writing of the manuscript and approved this submission.

To the best of our knowledge, this is the first report of a two-dimensional electrochemical-thermal model for a 38,120 LiFePO$_4$ LIB. Despite many works reported for battery separator, the effects of separator on *commercial* cylindrical LIBs have not been studied before. This paper numerically studies the effects of separator design on the performance of the commercial LIBs. The results herein provide quantitative, instead of qualitative, analyses of the subject.

We believe our work would appeal to the readership of *Energy & Fuels* in the areas of LIB and battery separator. We also declare no competing financial interest. In addition, the work present herein is original, unpublished, and not being considered for publication elsewhere.

Thank you for your time, and we look forward to hearing from you soon!

Sincerely,

Zhongchao Tan
Zhongchao Tan.

A typical cover letter is composed of the following elements in the order of their appearance:
1.   Writer's information (name, title, and address)
2.   Date (optional)
3.   Salutation
4.   Introduction
5.   Body
6.   Conclusion

7.   Complimentary close
8.   Signature
9.   Typed name

17.2.2 Date in Letter

Date is optional on your cover letter because it is considered the same date of the manuscript submission. Dates were necessary in the old times when manuscripts were delivered by mail, but now manuscripts are submitted online.

Date, if available, should be put on top of the letter instead of after the signature, which is the case in some countries. In addition, you need to make sure that date is formatted properly. The best format is *month day, year*. See Sect. 16.2.12.

17.2.3 Salutation

Salutation shows the formality of your writing. The tone of a cover letter should be polite and respectful. Always use formal language if you can. Informal language may sound less serious.

There are many choices for salutation. A common salutation begins with *Dear* (formally) or *To* (informally), followed by title and last name of the addressee, and ends with a comma. For example:

- Dear Professor Tan,
- Dear Dr. Tan,
- Dear Mr. Tan,
- To Dr. Tan,

You should use the highest level of title of the addressee. For example, a faculty member may prefer *Dr.* over *Mr.* or *Mrs.* In some countries, *Professor* is considered higher than Dr., but otherwise in other countries where someone can be a professor without a doctoral degree.

When you are writing to an institution without addressing a specific individual, you can use the following salutations.

- Dear Sir or Madam,
- To Whom It May Concern,
- Dear Editor,

Note

Hello or *Hi* is often used in an informal letter or email rather than a formal cover letter.

17.2.4 Introduction, Body, and Conclusion

You begin the body of your cover letter with a good purpose – *why* you are writing and *what* you are writing about. The purpose of a cover letter is to capture editor's attention.

The body of a cover letter introduces the *scope*, *novelty*, *importance*, and *impact* of your work. You also can mention the values to the target readers. A concise body constructed with simple sentences help get to your point directly. Nonetheless, being brief and concise should not scarify clarity; you need to provide all the information prescribed in the author's guideline.

The conclusion for a cover letter is a courtesy note, and it shows your expectation from the addressee. It normally consists of two sentences: the first is to show appreciation (e.g., *Thank you for your time*), and the second is your expectation (e.g., *we look forward to hearing from you soon*).

17.2.5 Complementary Close

A complimentary close is also referred to as a complimentary closing or signoff. A complimentary close can be a word or a phrase. It is above the sender's signature. The complimentary close always ends with a comma. For example:

- Sincerely,
- Sincerely yours,
- Best wishes,
- Regards,
- Best regards,
- Warm regards,
- With warmest regards,

17.3   Response to Reviewers' Comments

Another important communication between you and the editor handling your manuscript is *Response to Reviewers' Comments*. Most publishers use double-blind peer review. Thus, correspondence between you and the reviewers must go through the editor and the editorial office.

Despite the effort you put into your manuscript, peer reviewers always provide their comments from their point of view. Some comments are positive or complimentary, and the others may be critical. Occasionally, comments can be biased and unfriendly.

Defending your research requires communication skills and takes continual practice. Regardless of comments, they always strengthen your research and improve the quality of your manuscript. You ought to avoid emotional response. Always stay calm, professional, and appreciative when you respond to reviewers' comments. Professional responses with a positive attitude help you receive a positive decision.

Upon receival of the comments, you should carefully read through the reviewers' comments and understand each sentence. The editor normally gives you a month to resubmit the revised manuscript together with a separate file named *Response to Reviewer's Comments*. Thus, you have enough time to address their comments and concerns. Take your time and be patient in your response.

Within 1 week, but not immediately, you begin preparing for *Response to Reviewers' Comments*. Again, you indirectly correspond with the reviewers through the editorial office. You can start your response with a *thank you note* – thank the reviewers for their efforts put on your manuscript. This note shows the reviewers your politeness and appreciation. Example 17.2 consists of typical messages in this type of *note*. You also can convert this thank you note to a cover letter accompanying your resubmission. See also Sect. 17.1 Peer Review Process.

Example 17.2 Thank You Note

We sincerely thank the editor for the opportunity to improve and resubmit our manuscript; we also heartily thank the reviewers for their time spent reviewing the manuscript, the opportunity to address their concerns, and the comments to improve the manuscript prior to publication. We have taken their comments into serious consideration to improve the quality of our research. The revised manuscript tracks all changes, and the following is our responses to the comments.

Next, you respond to the reviewers' comments one by one, point by point. Meanwhile, you highlight or track all changes in the revised manuscript. An organized *Responses to Reviewers' Comments* and revised manuscript tracking all changes make easier for the editor and the reviewers to re-evaluate and reconsider your manuscript.

You can group the comments into four classes:

1. Positive.
2. Editorial.
3. Challenging.
4. Critical.

You can start with Classes 1 (positive) and 2 (editorial) before working on Classes 3 (challenging) and 4 (critical). In addition, your response depends on the comments.

Note

Use different colors to enhance visual clarity by distinguishing reviewers' comments and your responses. I normally chose black for reviewers' comments and blue for responses. Avoid red in this document. *See also* Sect. 14.6 Color for proper use of color.

17.3.1 Class 1 – Complimentary Comments

Example 17.3 is a complimentary comment. The reviewer points out that the manuscript has some limitations, but they are not listed. The reviewer probably accepted the manuscript *as is*.

You do not need to respond to complimentary and positive comments (Class 1). Many reviewers give positive feedback before negative ones, but they would not expect your responses to their complimentary comments.

Example 17.3 Complimentary Comments

This is a clear, well-prepared paper of high interest. The study has some limits, which however were duly presented and discussed by the authors. Overall, the manuscript provides a complete description of the study. Overall, I recommend this manuscript for publication in the present form.

However, you should keep positive comments in the file *Response to Reviewers' Comments*. The editor may share your file with all reviewers. Thus, all reviewers will see and consider these positive comments when they review your revised manuscript.

For another example, Reviewer #1 Comments in Appendix A4 – Sample Response to Reviewers' Comments – are all complimentary and positive. We did not respond to this reviewer's comments, but we kept them in the file.

17.3.2 Class 2 – Editorial Comments

Non-native English speakers often receive editorial comments. Example 17.4 shows three editorial comments listed in Appendix A4. The reviewer's comments certainly help improve the quality of writing. Meanwhile, they are easy to respond to; simply accept and revise accordingly. In addition, use line and page numbers to point the reviewers to the revisions in the revised manuscript.

17.3.3 Class 3 – Challenging Comments

Example 17.5 shows challenging but neutral comments (*see* Appendix A4). The reviewer asks two fair questions because the original manuscript lacks clarity – we should have justified the parameters chosen (*see* Chap. 6).

Example 17.4 Editorial Comments

P1L22 - it should be "Filtration of nanoparticles…" instead of "Filtration of nanoparticle". (*P1L22 means Paragraph 1; Line 22.*)

- **Response**: All phrases of "filtration of nanoparticle" are replaced with "filtration of nanoparticles" in the revised manuscript.

P2 and following – Authors may wish to divide the "Materials and Experimental Methods" section into smaller subsections, also the description of PVA nonwovens could be moved in front of the description of the experimental setup as in the section title.

- **Response:** Section 2 Materials and Experimental Methods is divided into two subsections, 2.1 Materials and 2.2 Experimental Methods, in the revised manuscript. See Page 4, Lines 64 and 78.

General remark – Authors may wish to use consistent unit system; the pressure drop is given in inches of H_2O, while the majority of other units are from the SI system.

- **Response:** SI units are used throughout the revised manuscript.

Our response begins with an explanation. Then, we answer the questions by specifying the revisions in the revised manuscript. Additionally, we copy and paste the updated text and point the reviewer to the page and line numbers in the revised manuscript. At the end, we list the reference (#21) used to address this comment. Such an organization allows the editor and the reviewers to easily see our responses rather than reading the entire manuscript again.

Example 17.5 Challenging Comments

P2L68–73 – Why such flow rates were chosen? Is the choice of the flow rate related to some standard requirements concerning testing of filtering media? Consider using the same units in case of the aerosol flow rate.

Response:

We chose the flow rate based on the nanoaerosol generator used in our study. The unit, liter per hour ($l.h^{-1}$), is labeled on the tungsten oxide generator, and converting it to another unit may confuse some readers. The carrier air of this generator is in the range of 50–500 $l.h^{-1}$ and the diluting air 180–1800 $l.h^{-1}$. However, in this set of experiment, we adjusted the flow rate to the range mentioned in the manuscript to generate three different size distributions for filtration tests [21].

The following text is added to the revised manuscript to address this comment (*see* Page 6, Lines 88–92):

"Since diffusion loss might be significant for such small particles, particle concentrations have to be large enough to be sufficient when the aerosol flow reaches the downstream measurement device. To detect these small particles, the aerosol flow rates were relatively high (4 lpm). A high flow rate is expected to shorten the travelling time of nanoparticles in the tubes and to reduce the diffusion loss of nanoparticles."

Reference: [21] Erickson, K.; Singh, M.; Osmondson, B. Measuring Nanoparticle Size Distributions in Real-Time: Key Factors for Accuracy. 2007.

Note

Editors and reviewers are busy people; be mindful in preparing *Responses to Reviewers' Comments*.

17.3.4 Class 4 – Critical Comments

Be mindful with the critical comments. Well-mannered communication makes a big difference. On the other hand, nothing is wrong about disagreeing with the reviewer. All you need is an evidence-based rebuttal with a convincing argument.

Example 17.6 is a response to critical comments. Despite our contributions to a database under five conditions, the reviewer criticizes that more data should be collected under other conditions. Our response starts with "We respectfully disagree with this comment for the following reasons." Then we give our reasons and the best practices in field experimental studies.

To conclude, you should always professionally communicate with the editorial office, the editor, and the reviewers with politeness. A cover letter and response to reviewers' comments are part of academic writing for engineering publications. They also require academic integrity, clarity, and conciseness. In addition, stay mindful, calm, confident, and polite when you respond to reviewers' comments. A carefully prepared response helps with a positive decision on your manuscript.

Good luck!

Example 17.6 Critical and Negative Comments

[**Comment 2**] The experiments were done without the mechanical ventilation which is not typical of a dental office. I suggest the experiment be repeated with the mechanical ventilation on to see the difference. It is very possible the air purifier is not necessary in dental offices if the ventilation is working, and it would help to see if the air purifier has an effect with the ventilation on.

- **Response:** We respectfully disagree with this comment for the following reasons. Our current research is focused on indoor air purification instead of mechanical ventilation system (*see* Lines 136–139 on Page 9). Although air circulation caused by mechanical ventilation decreases the removal rate of particles, it facilitates the distribution of contaminated air throughout spaces. In addition, the mechanical ventilation system in a real dental office was sensitive to the temperature and could not be adjusted; otherwise, it cannot represent real operating conditions. Therefore, it is not recommended in this study to create a relatively controlled environment for field test.

[**Comment 3**] The experiments should be repeated at different days to see if the results the authors obtained are reproducible.

- **Response:** We respectfully disagree with this comment for the following reasons. As much as we wish to repeat the tests on different days and in different dental offices, the costs and availability of partners prevent us from doing so. That is why field measurements are valuable to applied research like this one. In addition, we do not expect one single study to answer all technical questions in a field of study. We would have to leave the growth of database to other researchers.

Appendices

A1. Transitional Words and Phrases

(Source: Hacker and Sommers 2018)

To compare	also, alternatively, in comparison, in the same manner, likewise, similarly
To contrast	although, and yet, at the same time, but, despite, even though, however, in contrast, in spite of, nevertheless, on the contrary, on the other hand, still, though, yet
To give examples	for example, for instance, specifically, that is, to illustrate, as an illustration
To indicate logical relationship	accordingly, as a result, because, consequently, for this reason, hence, if, only if, otherwise, provided (that), providing (that), since, so, then, therefore, thus
To indicate sequence	first, second, third, initially, then, next, finally
To show addition	additionally, again, and, also, besides, equally important, further, furthermore, in addition, in the first place, moreover, next, too
To show place or direction	above, below, beyond, close, elsewhere, farther on, here, nearby, opposite, to the left (right, south, north, etc.)
To show time order	after, afterward, after that, as, as soon as, at last, before, before that time, during, earlier, finally, formerly, immediately, later, meanwhile, next, now, since, since then, shortly, subsequently, then, thereafter, until, when, while
To summarize or conclude	all in all, in conclusion, in other words, in short, in summary, on the whole, that is, therefore, thus, to sum up

---

A2. American and British English Spellings

(Sources: Oxford 2021; Tysto.com 2021; Williams 2018)

American and British English spellings are different, and either one of them should be used consistently. Admittedly, it is a challenge to non-native English speakers who speak neither of them to tell

© The Editor(s) (if applicable) and The Author(s), under exclusive license to Springer Nature Switzerland AG 2022
Z. Tan, *Academic Writing for Engineering Publications*, https://doi.org/10.1007/978-3-030-99364-1

the difference, but it is a mistake to mix them in one single document. The following sections briefly explain some common differences between American and British English.

Table A.1 shows some examples of American and British English spellings. A more comprehensive list of American and British spellings is available at Tysto.com (2021) website.

- -ed/−t

This difference lies in the past tense. American English only uses the *-ed* ending, while some past forms of verbs use *-t* instead of *-ed* in British English.

- −er/−re

Many words ending in *-er* in American English are spelled *-re* in British English. However, there are some exceptions. For example, *acre*, *massacre*, *mediocre*, and *ogre* are spelled the same.

- -l/−ll

Some words that have single consonant in American English require a double consonant with a suffix in British English. There are also exceptions: some words end with a single *l* in British English (e.g., fulfil) end with a double *l* in American English (e.g., fulfill).

- -og/−ogue

Nouns ending with *-og* (or *-ogue*) in American English may end with *-ogue* in British English.

- -or/−our

The cluster *-or* in American English is often spelled *-our* in British English, except for certain words like *your*, *yours*, *our*, etc.

- -z/−s

Most words that end in *-ize* or *-yze* in American English end in *-ise* or *-yse* in British English.

- Other common differences.

There are many minor differences between American and British English spellings. They include c/k, ou/o, ough/f, ough/w, ph/f, que/ck, q/ck, mme/m, y/i, etc.

Table A.1 Comparison of American and British spellings

Pair	American English	British English
-or/-our	Behavior	Behaviour
	Color	Colour
	Flavor	Flavour
	Honor	Honour
	Labor	Labour
-er/-re	Caliber	Calibre
	Center	Centre
	Fiber	Fibre
	Liter	Litre
	Maneuver	Manoeuvre
	Meter	Metre
	Theater	Theatre
	Specter	Spectre
-og/-ogue	Analog/analogue	Analogue
	Catalog/catalogue	Catalogue
	Dialog/dialogue	Dialogue
	Prolog/prologue	Prologue
-z/-s	Analyze	Analyse
	Criticize	Criticise
	Equalize	Equalise
	Recognize	Recognise
	Standardize	Standardise
-l/-ll	Canceled	Cancelled
	Counselor	Counsellor
	Equaled	Equalled
	Fueling	Fuelling
	Jeweler	Jeweller
	Marvelous	Marvellous
	Modeling	Modelling
	Traveling	Travelling
-ll/-l	Apall	Appal
	Fulfill	Fulfil
-ed/-t	Burned	Burned/burnt
	Leaned	Leaned/leant
	Learned	Learned/learnt
	Smelled	Smelled/smelt
	Spelled	Spelled/spelt
	Spoiled	Spoiled/spoilt
-um/-ium	Aluminum	Aluminium
-k/-que	Check	Cheque
-ay/-ey	Gray	Grey
-augh/-ough	Naught	Nought
-w/-ough	Plow	Plough
-m/-mme	Program	Programme
-Sk/-Sc	Skeptic	Sceptic
-f/-ph	Sulfur	Sulphur
-i/-y	Tire	Tyre
-y/-ey	Whisky	Whiskey

A3. Sample Outline

(Sources: This outline is one of many versions for the preparation of the paper by Li and Tan (2019).)

Title: Effects of Separator on the Electrochemical and Thermal Performances of Lithium-Ion Battery: A Numerical Study

1   Introduction

- Lithium-ion battery (LIB) is an important energy storage and transmission technology in a modern society.
- The key elements in a LIB include a cathode, an anode, an electrolyte, and a separator.
- Numerous researchers have experimentally studied battery materials to improve LIB performances.
- Alternatively, numerical models have been developed over the past decades for the optimization of battery design parameters.
- On the other hand, earlier modeling works are mainly focused on battery cathodes and anodes.
- Meanwhile, many experimental studies were reported to develop superior separator for LIB.
- This paper reports a two-dimensional electrochemical-thermal coupled model for the impact of separator in a 38,120 cylindrical LiFePO4 LIB.

2   Model Development

2.1   Model Domain and Assumptions

- Figure 1 illustrates the cylindrical domain of the electrochemical-thermal model developed in this study.

Figure 1. Schematic graph of lithium-ion battery

2.2   Governing Equations

2.2.1   Electrochemical Kinetics

- The local charge transfer current density on the electrode particle surface is calculated using:

2.2.2   Mass Conversion

- The mass conversion of lithium ions in electrode particles is calculated using the Fick's law:
- The boundary conditions are:
- The mass conversion of lithium ions in the electrolyte is calculated according to the concentrated solution theory.
- The boundary conditions are:

2.2.3   Charge Conversion

- The charge conversion in solid electrode particles is governed by Ohm's law:
- The boundary conditions are listed as follows:
- The charge conversion in liquid electrolyte is expressed by:
- The corresponding boundary conditions are.

2.2.4   Energy Conversion

- Three different types of heat are generated during the LIB charge and discharge, namely, reaction heat, ohmic heat, and active polarization heat.
- The calculated heat generations were then applied as heat source to simulate the thermal behavior of a cylindrical LIB unit (Figure).
- The corresponding boundary condition:

2.2.5   Numerical Solution of Model Equations

- The electrochemical-thermal coupled model was solved using the COMSOL Multiphysics 5.2 software.
- Table 1 and Table 2 show the model parameters and the material properties used in this study, respectively.

Table 1 Modeling parameters.
Table 2 Thermal properties of LIB materials

3   Results and Discussion

3.1   Model Validation

- The model developed for 38,120 type LiFePO$_4$ LIB was validated by comparing the model with the experimental data reported in the literature.

Figure 2. Comparison of model with experimental data.

- The validated model was then used to simulate the LIB performances with different separator design parameters.

3.2   Effects of Separator Thickness and Porosity on Energy Density

- Figure 3 shows the effects of separator thickness and porosity on the energy density of a LIB discharged at 2C rate.

Figure 3. Effects of separator thickness and porosity on the energy density.

- Positive effects of the separator thickness on the battery energy density than that of the separator porosity.
- For coin cell LIBs, however, the high porosity separator, the higher discharge rate capabilities.
- This diminished advantage of 38,120 LIB with a high porosity separator is attributed to the heat generation in the 38,120 cylindrical battery, which is much higher than that in a coin cell battery.

3.3   Electrolyte Concentration Distribution

- Figure 4 shows the electrolyte concentration distribution across the single battery cell with different separator thicknesses and porosities.

Figure 4. Electrolyte concentration distribution LIB with different separator thicknesses and porosities at the end of 2C rate discharge.

- During the battery discharge, the lithium ions are first extracted from the anode electrode particles, followed by intercalation into the cathode electrode.
- We further quantified the correlations between the separator porosity and electrolyte concentration gradient.
- Figure 5 presents the temporal changes of electrolyte concentration at separator boundaries.

Figure 5. Change of electrolyte concentration vs. separator porosity.

3.4   Effect of Separator on Thermal Behavior

- The heat generated in a LIB cell can be reversible heat or irreversible heat.
- Figure 6 shows the effects of the porosity and thickness of a separator on the heat generation rate within the separator.

Figure 6. Effect of porosity and length on the heat generation rate in the separator.

- The electrolyte concentration gradient decreases as the separator porosity increases.
- Figure 7 compares the effects of separator porosity and thickness on the average temperature rise in the packed material inside LIB.

Figure 7. Effect of separator porosity and length on the temperature rise.

- The excessive heat accumulation and uneven temperature distribution of LIB may degrade the LIB performance.
- Figure 8 shows the effects of separator thermal conductivity and heat capacity on the temperature rises and differences of the packed materials in LIBs.
- Figure 10. Effects of separator thermal conductivity and heat capacity on the temperature rises and differences of packed material in the battery cell discharged at 2C rate.
- The preceding analyses show that battery thermal performance can be improved by increasing both separator thermal conductivity and heat capacity.

A4. Sample Response to Reviewers' Comments

[Source: The file is for R Givehchi, Q Li, Z Tan. 2018. Filtration of sub-3.3 nm Tungsten oxide particles using nanofibrous filters. *Materials* 11 (8): 1277; https://doi.org/10.3390/ma11081277]

Response to Reviewers' Comments

We cordially thank the reviewers for their time and comments on the manuscript. The manuscript has been carefully revised and changed to a technical note by considering the comments and the suggestions to improve the quality of the work. The revisions have been presented in red color in the revised manuscript, and detailed responses to the reviewers are as follows.

Reviewer #1

In the manuscript materials – 328,023, the authors investigated the filtration efficiencies of triple modal tungsten oxide (WOx) airborne nanoparticles (0.82–3.3 nm in diameter) at three different concentrations. All tests were conducted using polyvinyl alcohol (PVA) nanofibrous filters at air relative humidity of 2.9%. Results outlined that the filtration efficiencies of sub-3.3 nm nanoparticles depended on the upstream particle concentration, and that the lower the particle concentration, the higher the filtration efficiency was.

This is a clear, well-prepared paper of high interest. The study has some limits, which however were duly presented and discussed by the authors. Overall, the manuscript provides a complete description of the study. Overall, I recommend this manuscript for publication in the present form.

Reviewer #2

In my opinion the topic of the paper has scientific and practical value and it relates to the scope of materials. The introduction is appealing and it gives the reader good generalized background information concerning nanofiltration. The aim of the study is also clearly stated and it encourages to read the rest of publication. The experimental design is described exhaustively and is appropriate to investigate the issues considered within the study. The results are discussed in detail and (when possible) compared with the results of already published literature. Few small concerns that need explaining are listed below along with editorial suggestions:

P1L22 – It should be "filtration of nanoparticles…" instead of "filtration of nanoparticle".

Response: All phrases of "filtration of nanoparticle" are replaced with "filtration of nanoparticles" in the revised manuscript.

P2 and following – Authors may wish to divide the "Materials and Experimental Methods" section into smaller subsections, also the description of PVA nonwovens could be moved in front of the description of the experimental setup as in the section title.

Response: Section 2. Materials and Experimental Methods is divided in two subsections, "2.1 Materials" and "2.2 Experimental Methods," in the revised manuscript. See Page 4, Lines 64 and 78.

P2L68–73 Why such flow rates were chosen? Is the choice of the flow rate related to some standard requirements concerning testing of filtering media? Consider using the same units in case of the aerosol flow rate.

Response: We chose the flow rate based on the nanoaerosol generator used in this study. The unit was chosen as liter per hour $(l.h^{-1})$ because it is the unit labeled on the tungsten oxide generator and converting it to another unit may confuse some readers. The carrier air of this generator is in the range of 50–500 $l.h^{-1}$

and the diluting air 180–1800 l.h^{-1}. However, in this set of experiments, the flow rate was adjusted in the mentioned range in the paper to generate three different size distributions for the filtration tests.

The following text is added to the revised manuscript to address this comment (see Page 6, Lines 88–92):

> Since diffusion loss might be significant for such small particles, particle concentrations have to be large enough to be sufficient when the aerosol flow reaches the downstream measurement device. To detect these small particles, the aerosol flow rates were relatively high (4 lpm). A high flow rate is expected to shorten the travelling time of nanoparticles in the tubes and to reduce the diffusion loss of nanoparticles [21].

Reference: [21] Erickson, K.; Singh, M.; Osmondson, B. Measuring Nanoparticle Size Distributions in Real-Time: Key Factors for Accuracy. 2007.

P3L115–116 – the first sentence of the paragraph is not necessary, please consider deleting it.

Response: This sentence is removed from the revised manuscript.

P6L176–181 – the relation between the descried result, i.e., "results herein showed obvious drops in filtration efficiencies as the size of nanoparticles decreased (even with the same upstream particle concentration)," and the conclusion, i.e., "conventional filtration model may need modification…the effect of particle concentration may have to be introduced into the models," is not clear. Authors may wish to elaborate to make the conclusion straightforward.

Response: This paragraph serves as a transition to the next topic, the effects of particle concentration on the filtration efficiency of nanoparticles, which is elaborated in the following paragraph. Nonetheless, to avoid confusion, this sentence is revised into:

> However, results herein showed obvious drops in filtration efficiencies as the size of nanoparticles were below 1.96 nm (see Page11, Lines187–188).

P7L184–188 – the most interesting part of the distribution is cut out from the investigation; authors say that particles larger than 1.96 are likely to drop below the detection limit. It's a pity that even partial results weren't discussed. Authors may wish to reconsider it. Perhaps some results for 1:10 dilution could be presented and discussed.

Response: As explained in *Experimental Methods*, the original concentrations of larger particles were in the range of 10^6 particles/cm^3. Excessive dilution would result in concentrations below the detection limit of the monitor. That is why we chose only three different orders of magnitude for concentrations (10^8 to 10^7 and 10^6 particles/cm^3). It did not go below 10^6. As such we could only discuss over smaller particles because.

The manuscript is revised as follows (see Page 11, Lines 194–198):

> Since the original concentrations of particles larger than 1.96 nm in diameter were in the order of 10^6 particles/cm^3 (see Fig.),the concentrations of large particles after dilution approached the detection limit of the FCE. Therefore, the effects of particle concentration on filtration efficiency are only presented herein for sub-1.8 nm nanoparticles.

General remark – authors may wish to use consistent unit system; the pressure drop is given is inches of H$_2$O, while the majority of other units are from the SI system.

Response: SI units are used throughout in the revised manuscript.

References

Alred GJ, Brusaw CT, Oliu WE. 2018. *Handbook of Technical Writing*, 12th edition, Bedford/st Martins, Boston, MA, USA.

APA (American Phycological Association). 2020. *Publication Manual of the American Psychological Association*, Seventh Edition. Washington, DC, USA. (https://www.apa.org/)

Bai Y, Yang K, Sun D, Zhang Y, Gao X. 2013. Numerical aerodynamic analysis of bluff bodies at a high Reynolds number with three- dimensional CFD modeling, *Science China Physics* 56: 277–289.

Cheng L. 2010. *Lignin Degradation and Dilute Acid Pretreatment for Cellulosic Alcohols Production*. MSc thesis, University of Cincinnati, Ohio, USA.

Dolan R, Yin S, Tan Z. 2010. Effect of headspace fraction and aqueous alkalinity on subcritical hydrothermal gasification of cellulose. *International Journal of Hydrogen Energy* 35: 6600-6610.

Du W, Bao X, Xu J, Wei W. 2006. Computational fluid dynamics (CFD) modeling of spouted bed: Assessment of drag coefficient correlations, *Chemical Eng Science* 61 (5): 1347-1740.

EPA (Environmental Protection Agency, USA). 2020. *2019 Automotive Trends Report: Greenhouse Gas Emissions, Fuel Economy, and Technology Since 1975*. Report number: EPA-420-R-20-006 (March 2020).

Fino D, Bensaid S, Piumetti M, Russo N, 2016. A review on the catalytic combustion of soot in Diesel particulate filters for automotive applications: From powder catalysts to structured reactors. *Applied Catalysis A: General* 509: 75-96.

Fuller S, Zhao Y, Cliff S, Wexler A, Kalberer M, 2012. Direct surface analysis of time-resolved aerosol impactor samples with ultrahigh-resolution mass spectrometry, *Analytical Chemistry* 84 (22): 9858-9864.

Ge H, Hu D, Li X, Tian Y, Chen Z, Zhu Y. 2015. Removal of low- concentration benzene in indoor air with plasma-MnO_2 catalysis system, *Journal of Electrostatics* 76: 216-221. (https://doi.org/10.1016/j.elstat.2015.06.003)

Givehchi R. 2015. *Filtration of NaCl and WOx Nanoparticles using Metal Wire Screen and Nanofibrous Filters*. PhD dissertation, University of Waterloo, Canada. (https://uwspace.uwaterloo.ca/handle/10012/10289)

Givehchi R, Du B, Razavi M, Tan Z, Siegel J. 2021. Performance of nanofibrous media in portable air cleaners, Aerosol Science and Technology (in press as of April 3, 2021). https://doi.org/10.1080/02786826.2021.1901846

Givehchi R, Li Q, Tan Z. 2018. Filtration of sub-3.3 nm Tungsten oxide particles using nanofibrous filters. Materials 11 (8): 1277; https://doi.org/10.3390/ma11081277

Hacker D, Sommers N. 2018. A Writer's Reference, ninth edition. Bedford/st Martins, New York, NY, USA.

Hagen R, Golombisky K. 2013. *White Space is Not Your Enemy: A Beginner's Guide to Communicating Visually Through Graphic, Web & Multimedia Design,* Taylor & Francis Group, New York, USA.

IEEE (Institute of Electrical and Electronics Engineers). 2020. *IEEE Editorial Style Manual for Authors* (V 04.10.2020). IEEE Publishing Operations, NJ, USA.

Imhoff R, Tanner R, Valente R, Luria M. 2000. The evolution of particles in the plume from a large coal-fired boiler with flue gas desulfurization, *Journal of the Air & Waste Management Association* 50: 1207-1214.

Khodabakhshi M, Wen J, Tan Z. 2021. Coefficient of restitution for silver nanoparticles colliding on a wet silver substrate, *Applied Surface Science* 554: 149607.

Li Y, Tan, Z. 2020. Effects of a separator on the electrochemical and thermal performances of lithium-ion batteries: a numerical study, *Energy & Fuels* 34 (11): 14915-14923.

Li Y, Li Q, Tan Z. 2019. A review of electrospun nanofiber-based separators for rechargeable lithium-ion batteries, *Journal of Power Sources* 443: 227262.

Z. Tan, *Academic Writing for Engineering Publications*, https://doi.org/10.1007/978-3-030-99364-1

Liu J, Nie Q, Tan Z, Luo Y, Wang S, Yu H. 2020. Insights into the impurities of Bi2WO6 synthesized using the hydrothermal method, *RSC Advances* 10: 40597–40607.

Liu C, Wang C, Kei C, Hsueh Y, Perng T. 2009. Atomic layer deposition of platinum nanoparticles on carbon nanotubes for application in proton-exchange membrane fuel cells, *Small* onlinelibrary.wiley.com/doi/abs/10.1002/smll.200900278

Min J. 2015. Quantifying the Effects of Winter Weather and Road Maintenance on Emissions and Fuel Consumptions. Master's Thesis, University of Waterloo, Canada.

MLA (Modern Language Association). 2020. *MLA Handbook*, eighth edition. MLA, New York, NY, USA (mlahandbook.org)

Munson B, Young D, Okiishi T, Huebsch W. 2009. *Fundamentals of Fluid Mechanics*, 6th ed. Wiley, New Jersey, USA.

Oxford. 2021. Oxford International English School Website accessed in January 2021. https://www.oxfordinternationalenglish.com/differences-in-british-and-american-spelling/

Pan W, Guo R, Zhang X, Ren J, Jin Q, Xu H. 2013. Absorption of NO by using aqueous $KMnO_4$ /$(NH_4)_2CO_3$ solutions, *Environmental Progress & Sustainable Energy* 32: 564–568.

Tan Z. 2014. *Air Pollution and Greenhouse Gases: from Basic Concepts to Engineering Applications for Air Emission Control*. Springer Verlag, Singapore.

The University of Chicago Press Editorial Staff. 2017. *The Chicago Manual of Style*, 17th edition. University of Chicago Press, Chicago, USA. (DOI: https://doi.org/10.7208/cmos17)

Thiel D. 2014. *Research Methods for Engineers*, Cambridge University Press. https://doi.org/10.1017/CBO9781139542326

Tysto. 2021. Comprehensive list of American and British spelling differences. (Online accessed January 2021. http://www.tysto. com/uk-us-spelling-list.html)

Wikipedia. 2021. List of scientists whose names are used as units. https://en.wikipedia.org/ Accessed March 2021.

Williams P. 2018. *Advanced Writing Skills for Students of English*. English Lessons Brighton. Amazon.Ca (Self publishing).

Yin S. 2011. Hydrothermal Conversion of Biomass to Bio-oil. PhD Thesis, University of Calgary, Canada.

Yu H. 2013. Absorption of nitric oxide from flue gas using ammoniacal cobalt (II) solutions, PhD Thesis, University of Waterloo, Canada. The others are not needed as they are part of the original text used as an example or practice problem.

Zhang B, Guo G, Zhu C, Ji Z, Lin C-H. 2020. Transport and trajectory of cough-induced bimodal aerosol in an air-conditioned space, *Indoor and Built Environment* 2020 Jul 22: 1420326X20941166.

Zhao B, Chen C, Tan Z. 2009. Modeling of ultrafine particle dispersion in indoor environments with an improved drift flux model, *Aerosol Science* 40: 29-43.

Zhou B, Zhao B, Tan Z. 2011. How particle resuspension from inner surfaces of ventilation ducts affects indoor air quality-a modeling analysis, *Aerosol Science and Technology* 45: 996-1009.

Zhou B, Zhao B, Guo X, Chen R, Kan H. 2013. Investigating the geographical heterogeneity in PM10-mortality associations in the China Air Pollution and Health Effects Study (CAPES): A potential role of indoor exposure to PM10 of outdoor origin, *Atmospheric Environment* 75: 217-223.

Index

Printed in Great Britain
by Amazon

44192316R00139